Winning Ways for Your Mathematical Plays, Volume 2

Winning Ways
for Your Mathematical Plays

Volume 2, Second Edition

Elwyn R. Berlekamp, John H. Conway, Richard K. Guy

A K Peters
Natick, Massachusetts

Editorial, Sales, and Customer Service Office

A K Peters, Ltd.
63 South Avenue
Natick, MA 01760

Library of Congress Cataloging-in-Publication Data

Berlekamp, Elwyn R.
Winning Ways for your mathematical plays / Elwyn Berlekamp, John H. Conway, Richard Guy.--2nd ed.
p. cm.
Includes bibliographical references and index.
ISBN 1-56881-130-6 (v. 1) – ISBN 1-56881-142-X (v. 2) – ISBN 1-56881-143-8 (v. 3) – ISBN 1-56881-144-6 (v. 4) (alk.paper)
1. Mathematical recreations. I. Conway, John Horton. II. Guy Richard K. III. Title.

QA95 .B446 2000
739.7'4--dc21 00-048541

Printed in Canada

07 06 05 04 03 10 9 8 7 6 5 4 3 2 1

To Martin Gardner

who has brought more mathematics to more millions than anyone else

Elwyn Berlekamp was born in Dover, Ohio, on September 6, 1940. He has been Professor of Mathematics and of Electrical Engineering/Computer Science at UC Berkeley since 1971. He has also been active in several technology business ventures. In addition to writing many journal articles and several books, Berlekamp also has 12 patented inventions, mostly dealing with algorithms for synchronization and error correction.

He is a member of the National Academy of Sciences, the National Academy of Engineering, and the American Academy of Arts and Sciences. From 1994 to 1998, he was chairman of the board of trustees of the Mathematical Sciences Research Institute (MSRI).

John H. Conway was born in Liverpool, England, on December 26, 1937. He is one of the preeminent theorists in the study of finite groups and the mathematical study of knots, and has written over 10 books and more than 140 journal articles.

Before joining Princeton University in 1986 as the John von Neumann Distinguished Professor of Mathematics, Conway served as professor of mathematics at Cambridge University, and remains an honorary fellow of Caius College. The recipient of many prizes in research and exposition, Conway is also widely known as the inventor of the Game of Life, a computer simulation of simple cellular "life," governed by remarkably simple rules.

Richard Guy was born in Nuneaton, England, on September 30, 1916. He has taught mathematics at many levels and in many places—England, Singapore, India, and Canada. Since 1965 he has been Professor of Mathematics at the University of Calgary, and is now Faculty Professor and Emeritus Professor. The university awarded him an Honorary Degree in 1991. He was Noyce Professor at Grinnell College in 2000.

He continues to climb mountains with his wife, Louise, and they have been patrons of the Association of Canadian Mountain Guides' Ball and recipients of the A. O. Wheeler award for Service to the Alpine Club of Canada.

Contents

Preface to the Second Edition

It's high time that there was a second edition of *Winning Ways*.

Largely as a result of the first edition, and of John Conway's *On Numbers and Games*, which we are glad to say is also reappearing, the subject of combinatorial games has burgeoned into a vast area, bringing together artificial intelligence experts, combinatorists, and computer scientists, as well as practitioners and theoreticians of particular games such as Go, Chess, Amazons and Konane: games much more interesting to play than the simple examples that we needed to introduce our theory.

Just as the subject of combinatorics was slow to be accepted by many "serious" mathematicians, so, even more slowly, is that of combinatorial games. But now it has achieved considerable maturity and is giving rise to an extensive literature, documented by Aviezri Fraenkel and exemplified by the book *Mathematical Go: Chilling Gets the Last Point* by Berlekamp and Wolfe. Games are fun to play and it's more fun the better you are at playing them.

The subject has become too big for us to do it justice even in the four-volume work that we now offer. So we've contented ourselves with a minimum of necessary changes to the original text (we are proud that our first formulations have so well withstood the test of time), with additions to the Extras at the ends of the chapters, and with the insertion of many references to guide the more serious student to further reading. And we've corrected some of the one hundred and sixty-three mistakes.

We are delighted that Alice and Klaus Peters have agreed to publish this second edition. Their great experience, and their competent and cooperative staff, notably Sarah Gillis and Kathryn Maier, have been invaluable assets during its production. And of course we are indebted to the rapidly growing band of people interested in the subject. If we mention one name we should mention a hundred; browse through the Index and the References at the end of each chapter. As a start, try *Games of No Chance*, the book of the workshop that we organized a few years ago, and look out for its successor, *More Games of No Chance*, documenting the workshop that took place earlier this year.

Elwyn Berlekamp, University of California, Berkeley
John Conway, Princeton University
Richard Guy, The University of Calgary, Canada

November 3, 2000

Preface

Does a book need a Preface? What more, after fifteen years of toil, do three talented authors have to add. We can reassure the bookstore browser, "Yes, this is just the book you want!" We can direct you, if you want to know quickly what's in the book, to the last pages of this preliminary material. This in turn directs you to Volume 1, Volume 2, Volume 3 and Volume 4.

We can supply the reviewer, faced with the task of ploughing through nearly a thousand information-packed pages, with some pithy criticisms by indicating the horns of the polylemma the book finds itself on. It is not an encyclopedia. It is encyclopedic, but there are still too many games missing for it to claim to be complete. It is not a book on recreational mathematics because there's too much serious mathematics in it. On the other hand, for us, as for our predecessors Rouse Ball, Dudeney, Martin Gardner, Kraitchik, Sam Loyd, Lucas, Tom O'Beirne and Fred. Schuh, mathematics itself is a recreation. It is not an undergraduate text, since the exercises are not set out in an orderly fashion, with the easy ones at the beginning. They are there though, and with the hundred and sixty-three mistakes we've left in, provide plenty of opportunity for reader participation. So don't just stand back and admire it, work of art though it is. It is not a graduate text, since it's too expensive and contains far more than any graduate student can be expected to learn. But it does carry you to the frontiers of research in combinatorial game theory and the many unsolved problems will stimulate further discoveries.

We thank Patrick Browne for our title. This exercised us for quite a time. One morning, while walking to the university, John and Richard came up with "Whose game?" but realized they couldn't spell it (there are three tooze in English) so it became a one-line joke on line one of the text. There isn't room to explain all the jokes, not even the fifty-nine private ones (each of our birthdays appears more than once in the book).

Omar started as a joke, but soon materialized as Kimberley King. Louise Guy also helped with proof-reading, but her greater contribution was the hospitality which enabled the three of us to work together on several occasions. Louise also did technical typing after many drafts had been made by Karen McDermid and Betty Teare.

Our thanks for many contributions to content may be measured by the number of names in the index. To do real justice would take too much space. Here's an abridged list of helpers: Richard Austin, Clive Bach, John Beasley, Aviezri Fraenkel, David Fremlin, Solomon Golomb, Steve Grantham, Mike Guy, Dean Hickerson, Hendrik Lenstra, Richard Nowakowski, Anne Scott, David Seal, John Selfridge, Cedric Smith and Steve Tschantz.

No small part of the reason for the assured success of the book is owed to the well-informed and sympathetic guidance of Len Cegielka and the willingness of the staff of Academic Press and of Page Bros. to adapt to the idiosyncrasies of the authors, who grasped every opportunity to modify grammar, strain semantics, pervert punctuation, alter orthography, tamper with traditional typography and commit outrageous puns and inside jokes.

Thanks also to the Isaak Walton Killam Foundation for Richard's Resident Fellowship at The University of Calgary during the compilation of a critical draft, and to the National (Science & Engineering) Research Council of Canada for a grant which enabled Elwyn and John to visit him more frequently than our widely scattered habitats would normally allow.

And thank you, Simon!

University of California, Berkeley, CA 94720 — *Elwyn Berlekamp*
University of Cambridge, England, CB2 1SB — *John H. Conway*
University of Calgary, Canada, T2N 1N4 — *Richard Guy*

November 1981

You are now here

If you want to know roughly what's elsewhere, turn to the little notes about our four main themes:

There are a number of other connexions between various chapters of the book:

THEORY (chapters 1–13) | PRACTICE (chapters 14–25)

General Addition (1–8) | Other Theories (9–13) | Case Studies (14–25)

♠ before 1; ♡ before 9; ♣ before 14; ◇ before 23

1 2 3 4 5 6 7 8 9 10 11 12 13 14 15 16 17 18 19 20 21 22 23 24 25

Numbers (1–2) | Nimbers (3–4) | Heat (5–6) | Hackenbush (7–8) | Simultaneous moves (9–10) | Loopy (11–12) | Coins (14–15) | Pencil&Paper (16–17) | Trapping (18–20) | On Your Own (23–25)

Nim-like Games (12–18)

Basics (1–3) | Varied Values (5–7) | Tabulating (9–18) | Board Games (19–24)

However, you should be able to pick any chapter and read almost all of it without reference to anything earlier, except perhaps the basic ideas at the start of the book.

Change of Heart!

New styles of architecture, a change of heart.
Wystan Hugh Auden, *Sir, no man's enemy.*

I have heard her declare, under the rose, that Hearts was her favourite suit.
Charles Lamb, *Essays of Elia*, Mrs. Battle's Opinions on Whist.

So far our compound games have been played by two players who move alternately in just one component at a time, and the rules have ensured that they always end, the last player to move being the winner. Now for a change of heart, let's see what happens when we break some of these rules.

In Chapter 9, you must move in *every* component, and in Chapter 10, you can move in whatever components you like.

In Chapter 11, there are some partizan games with infinitely many positions, and some other loopy games in which play might continue forever.

Chapter 12 deals with the rather different theory of impartial loopy games, and with some other modifications of the impartial theory, which might allow a player to make several consecutive moves.

Chapter 13 gives the theory of impartial games when the last player is declared to be the *loser.*

-9-

If You Can't Beat 'Em, Join 'Em!

Remote from towns he ran his godly race.
Oliver Goldsmith, *The Deserted Village*, l. 143.

This suspense is terrible. I hope it will last.
Oscar Wilde, *The Importance of Being Earnest*, III.

All the King's Horses

In *sums* of games it is a move in just *one* part that counts as a move in the sum. Now we consider the **join** of several games in which we must move in *every* part.

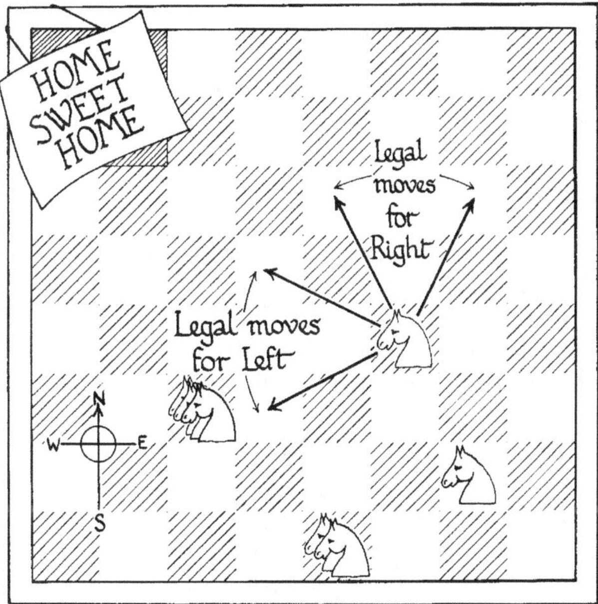

Figure 1. How Horses Head for Home.

We shall play our first few games on the 8 by 8 chessboard illustrated in Fig. 1 with a number of horses. As his move, Left must move *every* horse he can two places West and one place North or South; Right, for his part, must move every horse he can two places North and one place East or West, as in Fig. 1. So the horses move rather like knights in Chess, but there are several differences. Each player is limited to just 2 of the possibly 8 directions a knight can move in; there may be arbitrarily many horses on the same square; and the same horse is moved by both players (the horses belong to the King, not to Left or Right). Compare the White Knight in Chapter 3.

A player will be unable to move in the game if there is any one horse which he can't move. According to the *normal play rule* he would then lose, but in this chapter we shall also give equal treatment to the *misère play rule* under which he would win.

We Can Join Any Games

Our game can be regarded as made by joining together a number of one-horse games. In fact any games

$$G, \quad H, \quad K, \quad \ldots$$

can be played simultaneously like this to obtain a compound game

$$G \wedge H \wedge K \wedge \ldots \qquad (\text{“}G \text{ and } H \text{ and } K \text{ and} \ldots\text{”})$$

called the **conjunctive compound** by Smith and in ONAG, but here, for short, their **join**. To move in the compound game you must make a move legal for you in *every* one of the component games

$$G, \quad H, \quad K, \quad \ldots$$

rather than in just one, as in the *sum* or disjunctive compound. If you cannot do so you *lose* in normal play, but *win* in misère play.

How Remote Is a Horse?

Everything depends on the first horse to finish, for this stops the whole game. (In this game a horse finishes when the player whose turn it is cannot move him.) If one horse—the **favorite**—seems nearer to finishing than the others you should treat it with particular care. Move it so as to win quickly if you can, and otherwise so as to postpone defeat as long as possible in the hope of bringing up a more favorable horse to finish first—in short:

The maxim holds for joins of any games. When we know who starts, a game played in this way lasts for a perfectly definite number of moves which C.A.B. Smith has called its *Steinhaus function* or *remoteness.*

We use the term **left remoteness** when Left starts, and **right remoteness** when Right starts. Since the turns alternate, we need only consider the Right remotenesses of Left's options and the left remotenesses of Right's. You should try to leave an *even* remoteness (as small as possible) for your opponent, so as to ensure that when the remoteness is reduced to zero it will be his turn to move. This is because (in normal play)

$\mathcal{P}$-positions have even remoteness
$\mathcal{N}$-positions have odd remoteness

So remotenesses can be worked out by the rules:

For the *left* remoteness of G in
normal play, take
1 more than the LEAST EVEN *right* remoteness
of any G^L which has even right remoteness,
otherwise
1 more than the GREATEST ODD *right* remoteness
if all G^L have odd right remoteness,
and finally
0, if G has no left option.

Or, more concisely:

FOR NORMAL PLAY
Use
1 + LEAST EVEN
if possible,
1 + GREATEST ODD
if not, or
ZERO
if you've no option.

To find the *right* remoteness, use the *left* remotenesses of the *right* options, G^R, but still prefer LEAST EVEN else GREATEST ODD.

$R_L^+ R_R^+$ (normal play)

00	00	10	10	10	10	10	10
00	00	10	10	10	10	10	10
01	01	11	12	12	12	12	12
01	01	21	22	22	32	32	32
01	01	21	22	22	32	32	32
01	01	21	23	23	33	34	34
01	01	21	23	23	43	44	44
01	01	21	23	23	43	44	44

(a) First horse stuck loses.

$R_L^- R_R^-$ (misère play)

00	00	10	10	10	10	10	10
00	00	20	30	30	30	30	30
01	02	22	42	52	52	52	52
01	03	24	44	64	74	74	74
01	03	25	46	66	86	96	96
01	03	25	47	68	88	A8	B8
01	03	25	47	69	8A	AA	CA
01	03	25	47	69	8B	AC	BB

(b) First horse stuck wins.

$R_L^+ R_R^+$ (normal with pass)

00	00	12	12	34	34	56	56
00	00	12	12	34	34	56	56
21	21	11	14	34	36	56	56
21	21	41	44	44	56	56	76
43	43	43	44	44	56	56	76
43	43	63	65	65	55	58	76
65	65	65	65	65	85	88	86
65	65	65	67	67	67	68	66

(c) First horse home wins.

$R_L^- R_R^-$ (misère with pass)

00	00	12	12	45	56	56	56
00	00	23	34	34	34	67	78
21	32	22	42	42	56	56	56
21	43	24	44	54	64	74	77
54	43	24	45	66	76	86	76
65	43	65	46	67	66	98	A8
65	76	65	47	68	89	88	B9
65	87	65	77	67	8A	9B	AA

(d) First horse home loses.

R^+ (normal impartial)

0	0	1	1	2	2	3	3
0	0	1	1	2	2	3	3
1	1	1	1	3	3	3	3
1	1	1	3	3	3	3	5
2	2	3	3	4	4	5	5
2	2	3	3	4	4	5	5
3	3	3	3	5	5	5	5
3	3	3	5	5	5	5	6

(x) First horse home wins.

R^- (misère impartial)

0	0	1	1	3	4	3	3
0	0	2	3	2	2	4	5
1	2	2	2	2	4	4	4
1	3	2	3	4	5	6	5
3	2	2	4	5	4	5	6
4	2	4	5	4	7	6	7
3	4	4	6	5	6	7	6
3	5	4	5	6	7	6	7

(y) First horse home loses.

Table 1. How Remote Are All the Horses? (A=10, B=11, C=12.)

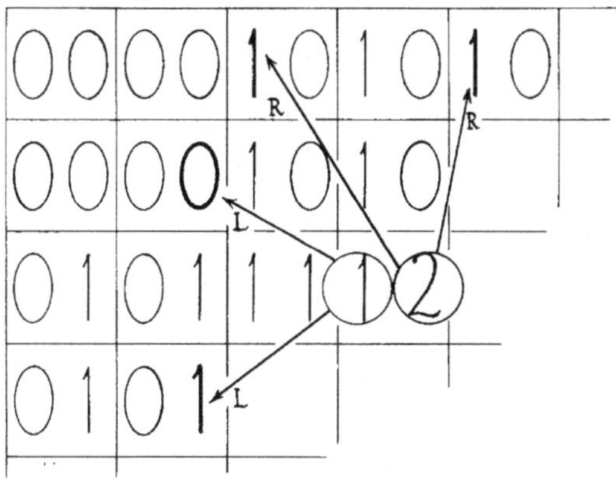

Figure 2. How Remote Is a Horse?

Table 1(a) gives the left and right remotenesses (in normal play) for horses in every possible position. See how a 0 on the left side corresponds to a position from which Left cannot move. Figure 2 illustrates a case with larger remotenesses. Here Left's two options have (right) remotenesses 0 and 1; of these he prefers the only even number, 0, and adds 1 to obtain 1. Right's two options have (left) remoteness 1 (not much choice!) and he adds 1 to obtain 2.

What If the First Horse to Get Stuck Wins?

In the *misère play* version, winner and loser are interchanged and so the players will prefer to move to *odd* rather than even. The rule for computing the misère play remotenesses is therefore, in condensed form:

FOR MISÈRE PLAY
Use
1 + LEAST ODD
if possible,
1 + GREATEST EVEN
if not, or
ZERO
if you've no option.

Table 1(b) gives the misère play remotenesses for our game.

Since, in either case, a join of games finishes when its *first* component does, its remoteness (of any kind) is the *least* remoteness (of the same kind) of any of the components:

$$R_L^+(G \wedge H \wedge \ldots) = \min(R_L^+(G), R_L^+(H), \ldots),$$
$$R_R^+(G \wedge H \wedge \ldots) = \min(R_R^+(G), R_R^+(H), \ldots),$$
$$R_L^-(G \wedge H \wedge \ldots) = \min(R_L^-(G), R_L^-(H), \ldots),$$
$$R_R^-(G \wedge H \wedge \ldots) = \min(R_R^-(G), R_R^-(H), \ldots).$$

In this and similar contexts we use:

L for Left starting,
R for Right starting,
$+$ for normal play,
$-$ for misère play.

To *win*, move to a position for which your *opponent*'s remoteness is EVEN in NORMAL play, ODD in MISÈRE play.

Let's see who wins

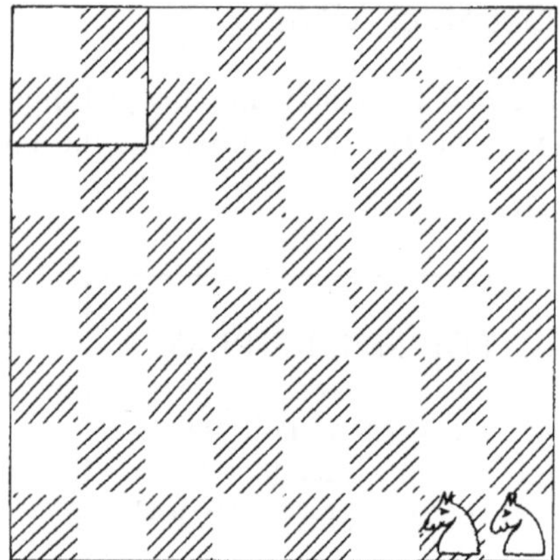

in MISÈRE play.

The left remotenesses of the two horses are 10 and 11 (A and B) so the left remoteness of the position as a whole is the least of these, 10. Since this is even, Left has a good move which changes this to the ODD number 9, but the right remotenesses of these horses are 12 and 11 (C and B), minimum 11, and so from this position Right has no good move. Left's favorite horse is the left one, but Right's is the right one, even though this seems further from finishing.

A Slightly Slower Join

We can get a more interesting game by changing the rules slightly. If one player cannot move some horse which the opponent could, we may allow him to make a *pass* move for that horse, but he must still make proper moves with all the horses he can. The game will now end as soon as the first horse reaches **home**, the top left 2 by 2 square, since then *neither* player can move this horse and passes are not allowed. The normal and misère remotenesses for this version are shown in Tables 1(c) and (d). They are calculated in exactly the same way, but taking account of the new pass moves. For the position whose misère remotenesses are being computed in Fig. 3, Left has a proper move to a position of right remoteness 3, so his remoteness is $1+3=4$. Right has no proper move, but has a pass move to the same position with Left to move, so his remoteness is $1+4=5$.

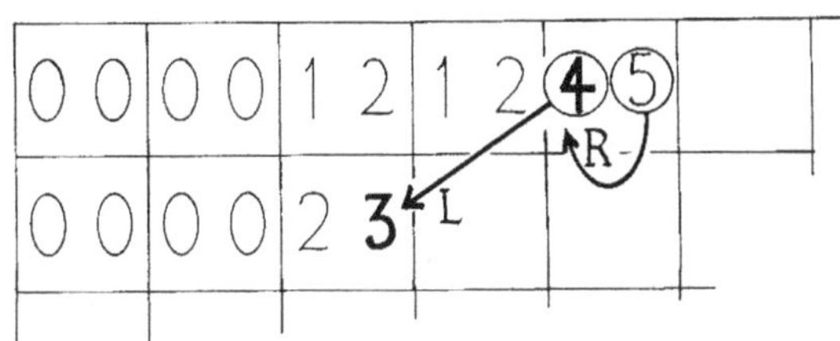

Figure 3. Right's a Bit More Remote Because He's Stuck.

Moving Horses Impartially

As a further variant, we may make moves in all the four directions of Fig. 1 legal for both players, so that the game becomes an impartial one, and there is no difference between the left and right remotenesses of any position.

In Tables 1(x) and 1(y) for the normal and misère versions there is therefore only one digit in each square.

For this game, the horse that seems to be ahead in

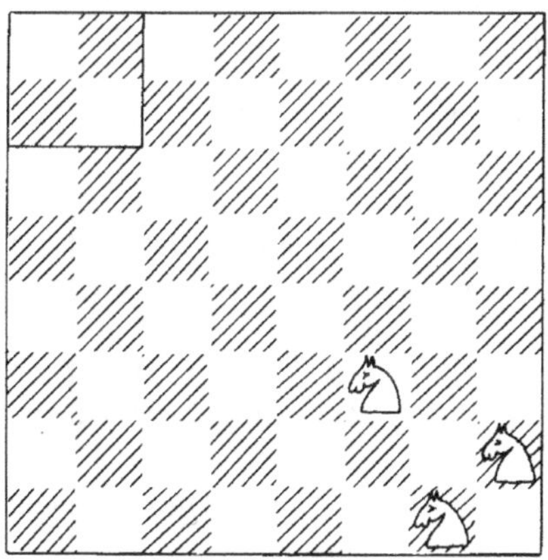

really *is* the favorite in normal play (remoteness 4 against two 5s), but the two trailing ones are joint favorites in misère play (remotenesses 6 as against 7).

All these games may be played on any size of board, or even on a quarter-infinite one. Table 5 (in the Extras) gives remotenesses for this latter case.

Cutting Every Cake

The game of Cutcake has a conjunctive version, **Cutcakes**, played with the same equipment (see Fig. 3 of Chapter 2), in which Lefty must make a vertical cut (or Rita a horizontal one), along a scored line, in *every* piece of cake. So the first player to produce the kind of strip in which his opponent has no legal move, wins in normal play, *loses* in misère. Tables 2(a) and 2(b) give the remotenesses, and Fig. 4 indicates how they were calculated.

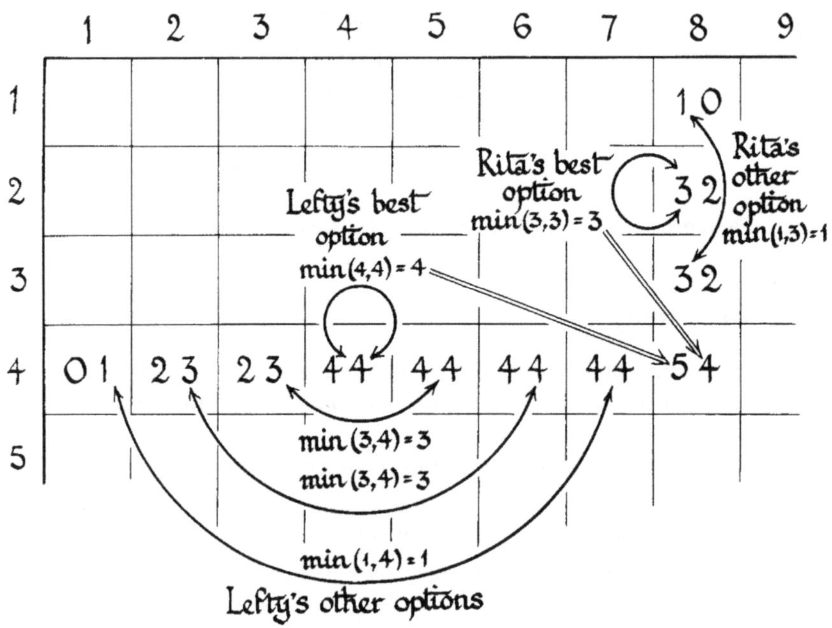

Figure 4. Lefty and Rita Ponder a 4 by 8 Cake.

In another version (Tables 2(c) and 2(d)) a player who cannot cut a particular piece of cake may pass over that piece provided his opponent can still cut it. The game ends when the *first* 1 by 1 cake appears, for this admits no cut by *either* player. So for a horizontal strip, which cannot be cut by Rita, her remoteness is one more than Lefty's.

In impartial Cutcakes (Tables 2(x) and 2(y)) the players must cut all the cakes, but each may do so in either direction.

In some of these tables we have "writ large" some remotenesses which are the same for the whole blocks of entries.

$R_L^+ R_R^+$ (normal play).

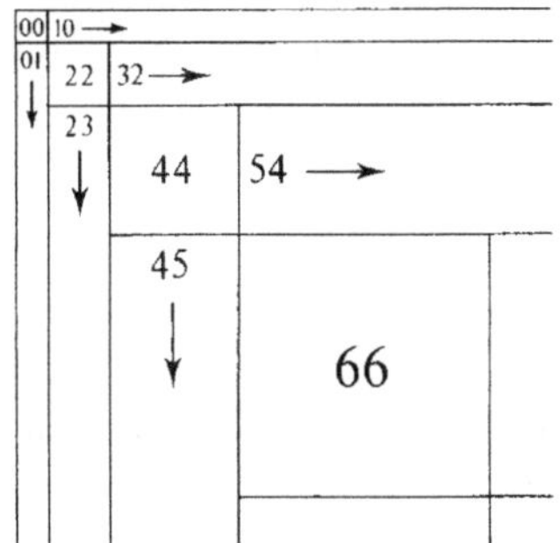

(a) First appropriate strip wins.

$R_L^- R_R^-$ (misère play).

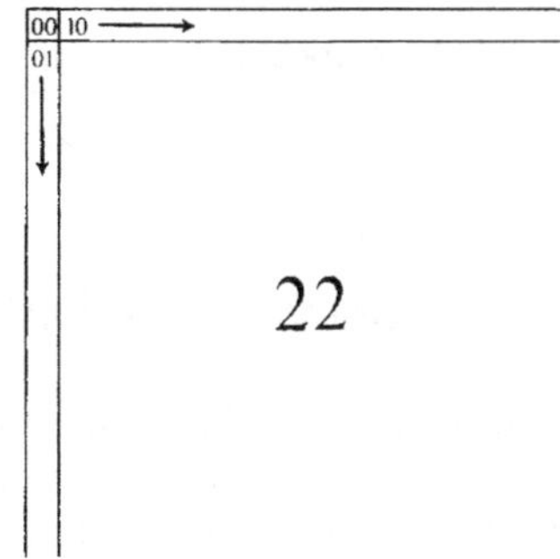

(b) First appropriate strip loses.

$R_L^+ R_R^+$ (normal with pass).

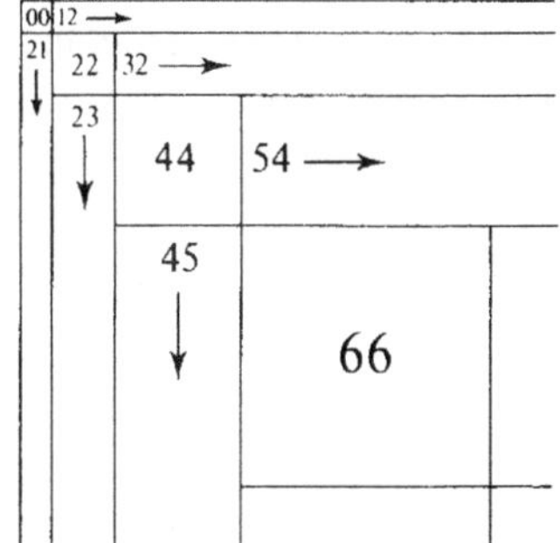

(c) First 1×1 cake wins.

$R_L^- R_R^-$ (misère with pass).

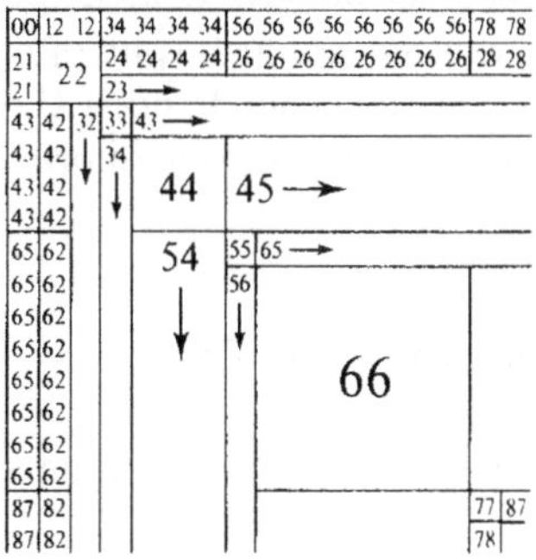

(d) First 1×1 cake loses.

R^+ (normal impartial).

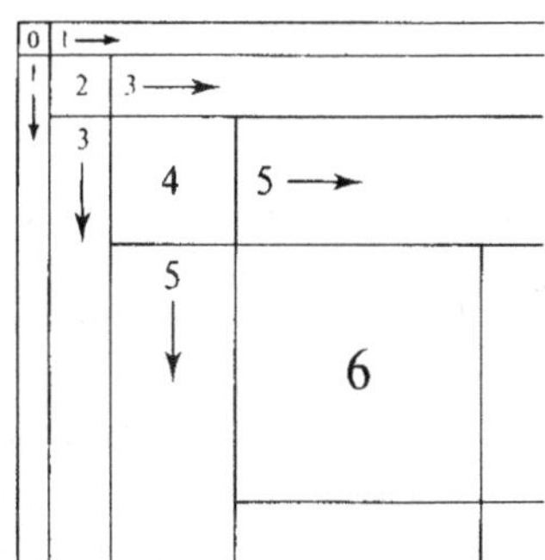

(x) First 1×1 cake wins.

R^- (misère impartial).

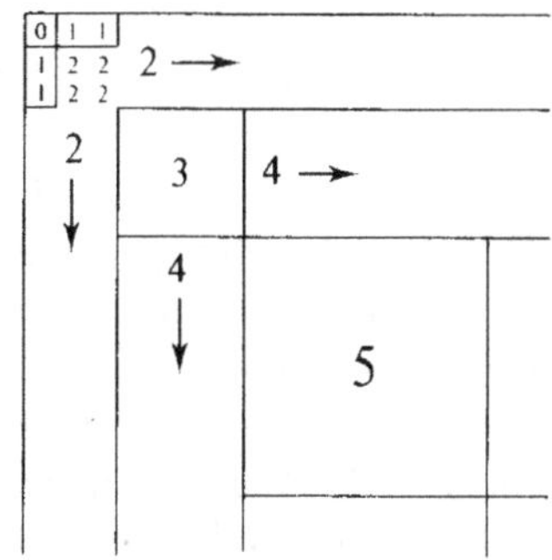

(y) First 1×1 cake loses.

Table 2. How Long It Takes to Cut Our Cakes.

Eatcakes

Remoteness tables for **Eatcakes**, a variant game which is more natural as a *slow* join (see later), are given in the Extras (Table 7).

When to Put Your Money on the Last Horse

In our next variation of All the King's Horses, an apparently trivial modification of the rules produces a dramatic change in tactics. Last time we allowed you to pass for a horse you couldn't move only if your opponent could still move that horse. Now we allow you to pass provided that there is *any* horse that one of you can still move, and the game only ends when *all* the horses reach home. In normal play whoever takes the last horse home is the winner; he is of course the loser in misère play.

In this game the race is not to the swift—it is the horse likely to be the *last* in the race (the **outsider**) who must be moved with special care. If you think that you can win with him you should try to win *slowly*, lest your opponent hold back another horse to make the race finish in *his* favor. On the other hand if the outsider looks like he's losing for you, take him home quickly and hope to leave some more promising laggard on the course.

Slow Horses Join the Also-Rans

Our game exemplifies a new way of joining several games,

$$G,\ H,\ K,\ \ldots$$

to produce a compound,

$$G \triangle H \triangle K \triangle \ldots \qquad (\text{“}G \text{ also } H \text{ also } K \text{ also } \ldots\text{”})$$

called the *continued conjunctive compound* in ONAG, and here, the **slow join**. Our previous kind of join may be called the *rapid join* when we need to avoid confusion. In the slow join of a number of games a player must move in every component he can, and the game ends only when he cannot move *anywhere.*

The best tactics are a travesty of those for the rapid join – move *slowly* when you're *winning*, *quickly* when you're *losing*! The winner is anxious to savor his inexorable superiority for as long as possible, while the loser wants to get it over with, but quick! Given who starts, a game lasts, when played according to these cat and mouse tactics, for a perfectly definite number of moves, called the **suspense number**.

We find these suspense numbers by a parody of the remoteness rules:

For the *left* suspense of G in
normal play, take
1 more than the GREATEST EVEN *right* suspense
of any G^L which has even right suspense,
otherwise
1 more than the LEAST ODD *right* suspense
if all G^L have odd right suspense,
and finally
0, if G has no left option.

In short:

FOR NORMAL PLAY
Use
1 + GREATEST EVEN
if possible,
1 + LEAST ODD
if not, or
ZERO
if you've no option.

And similarly:

FOR MISÈRE PLAY
Use
1 + GREATEST ODD
if possible,
1 + LEAST EVEN
if not, or
ZERO
if you've no option.

A handy maxim covering both cases is:

Since a slow join finishes only when its *last* component does, its suspense number (of any kind) is the *greatest* suspense number (of the same kind) of any of the components:

$$\begin{aligned} S_L^+(G\triangle H\triangle\ldots) &= \max(S_L^+(G), S_L^+(H),\ldots),\\ S_R^+(G\triangle H\triangle\ldots) &= \max(S_R^+(G), S_R^+(H),\ldots),\\ S_L^-(G\triangle H\triangle\ldots) &= \max(S_L^-(G), S_L^-(H),\ldots),\\ S_R^-(G\triangle H\triangle\ldots) &= \max(S_R^-(G), S_R^-(H),\ldots). \end{aligned}$$

$S_L^+ S_R^+$ (normal with pass)

00	00	12	12	34	34	56	56
00	00	12	12	34	34	56	56
21	21	11	12	32	34	54	56
21	21	21	22	22	34	34	56
43	43	23	22	22	34	34	56
43	43	43	43	43	33	34	54
65	65	45	43	43	43	44	44
65	65	65	65	65	45	44	44

(a) Last horse home wins.

$S_L^- S_R^-$ (misère with pass)

00	00	12	12	45	34	56	56
00	00	23	12	34	34	67	56
21	32	22	22	44	36	56	56
21	21	22	44	34	44	56	77
54	43	44	43	44	56	66	56
43	43	63	44	65	66	58	66
65	76	65	65	66	85	66	77
65	65	65	77	65	66	77	66

(b) Last horse home loses.

S^+ (normal impartial)

0	0	1	1	2	2	3	3
0	0	1	1	2	2	3	3
1	1	1	3	3	3	3	3
1	1	3	3	3	3	5	5
2	2	3	3	4	4	5	5
2	2	3	3	4	4	5	5
3	3	3	5	5	5	5	5
3	3	3	5	5	5	5	6

(x) Last horse home wins.

S^- (misère impartial)

0	0	1	1	3	2	3	3
0	0	2	1	2	2	4	3
1	2	2	4	2	4	4	4
1	1	4	3	2	3	6	5
3	2	2	2	5	4	3	4
2	2	4	3	4	3	6	5
3	4	4	6	3	6	5	6
3	3	4	5	4	5	6	7

(y) Last horse home loses.

Table 3. The Suspense of Slow Horse Racing.

Compare the two theories:

RAPID REMOTENESS MINIMUM (favorite)	as against	SLOW SUSPENSE MAXIMUM (outsider)

Tables 3(a) to 3(y) give suspense numbers for the four versions of All the King's Horses in which the *last* horse determines the race.

Tables 3(a) and 3(b) are for the normal and misère version we have already described in which you may make a pass move for any horse you cannot move properly.

Tables 3(x) and 3(y) are for the impartial versions of the game, when left and right suspense numbers coincide.

Let Them Eat Cake

Sometimes the way we play a game makes a rapid join more sensible than a slow one, or vice versa. For Cutcakes the rapid join was more natural; when we eat our cake it is less so. Tables 6 and 7, showing suspense numbers for Cutcakes and remotenesses for Eatcakes, have therefore been relegated to the Extras of this chapter.

In **Eatcakes** Lefty must eat away a vertical, or Rita a horizontal, strip of width 1 from every cake which is still on the table. This move will separate the cake into two pieces unless the eaten strip was along an edge. The last mouthful wins in normal play, loses in misère play. The disjunctive version of this game (due to Jim Bynum) was called Eatcake in Chapter 8.

Once again each cake defines a component game, but the whole game does not end when a cake is completely consumed, since nobody can see it any more. You don't have to eat a cake that isn't there! So the join is automatically a slow one—the normal and misère suspense numbers appear in Tables 4(a) and 4(b), and suspense numbers for the impartial version, in which the players eat strips in either direction, are displayed in Tables 4(x) and 4(y). For the impartial games the rows are ultimately periodic. Rows 0 through 9 of Table 4(x) have periods dividing 16, while for rows 0 through 4 of Table 4(y) the periods divide 18. In each case the first 20 entries contain a full period.

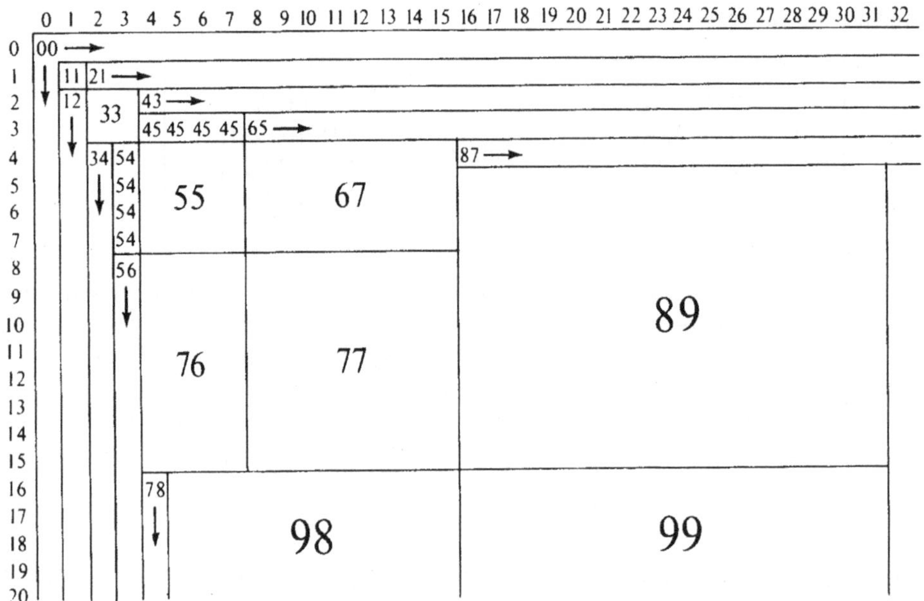

Table 4(a). Eating Cakes Normally ($S_L^+ S_R^+$).

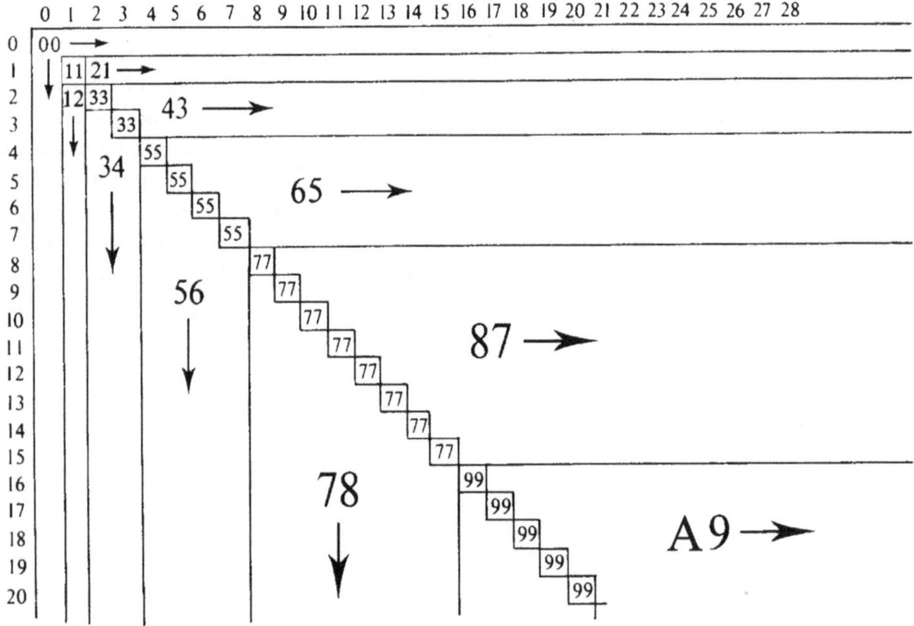

Table 4(b). Eating Cakes Miserably ($S_L^- S_R^-$). (A=10.)

	0	1	2	3	4	5	6	7	8	9	10	11	12	13	14	15	16	17	18	19	20	21	22	23	24	25	26	27	28	29	30	31
0	0	0	0	0	0	0	0	0	0	0	0	0	0	0	0	0	0	0	0	0	0	0	0	0	0	0	0	0	0	0	0	0
1	0	1	1	1	1	1	1	1	1	1	1	1	1	1	1	1	1	1	1	1	1	1	1	1	1	1	1	1	1	1	1	1
2	0	1	2	3	3	3	2	3	3	3	2	3	3	3	2	3	3	3	2	3	3	3	2	3	3	3	2	3	3	3	2	3
3	0	1	3	2	3	3	3	3	2	3	3	2	3	3	3	3	2	3	3	2	3	3	3	3	2	3	3	2	3	3	3	3
4	0	1	3	3	4	5	5	5	5	5	3	3	4	5	5	5	5	5	3	3	4	5	5	5	5	5	3	3	4	5	5	5
5	0	1	3	3	5	4	5	5	5	5	3	5	5	4	5	5	5	5	3	5	5	4	5	5	5	5	3	5	5	4	5	5
6	0	1	2	3	5	5	4	5	5	5	5	4	5	5	5	5	4	5	5	5	5	5	4	5	5	5	5	4	5	5	5	5
7	0	1	3	3	5	5	5	4	5	5	5	5	5	5	4	5	5	5	5	3	5	5	5	4	5	5	5	5	5	5	4	5
8	0	1	3	2	5	5	5	5	6	7	7	7	7	7	7	7	7	7	4	5	5	5	5	5	6	7	7	7	7	7	7	7
9	0	1	3	3	5	5	5	5	7	6	7	7	7	7	7	7	7	7	5	7	5	5	5	5	7	6	7	7	7	7	7	7
10	0	1	2	3	3	3	5	5	7	7	4	5	5	5	5	5	5	6	7	7	7	7	7	7	7	7	4	5	7	7	7	7
11	0	1	3	2	3	5	4	5	7	7	5	5	6	7	7	7	7	7	7	7	7	5	6	7	7	7	7	7	7	7	7	6
12	0	1	3	3	4	5	5	5	7	7	5	6	7	7	7	7	7	7	7	7	6	7	7	7	7	7	7	7	7	6	7	7
13	0	1	3	3	5	4	5	5	7	7	5	7	7	6	7	7	7	7	7	7	7	7	7	6	7	7	7	7	7	7	7	7
14	0	1	2	3	5	5	5	4	7	7	5	7	7	7	6	7	7	7	7	7	7	7	7	7	7	7	5	6	7	7	7	7
15	0	1	3	3	5	5	5	5	7	7	5	7	7	7	7	6	7	7	7	7	7	7	7	7	7	7	7	7	6	7	7	7
16	0	1	3	2	5	5	4	5	7	7	5	7	7	7	7	7	6	7	7	7	7	7	7	7	7	7	5	7	7	7	8	9
17	0	1	3	3	5	5	5	5	7	7	6	7	7	7	7	7	7	7	7	6	7	7	7	7	7	7	7	7	7	7	9	7
18	0	1	2	3	3	3	5	5	4	5	7	7	7	7	7	7	7	7	6	7	7	7	7	7	7	7	7	7	7	7	9	7
19	0	1	3	2	3	5	5	3	5	7	7	7	7	7	7	7	7	6	7	7	7	7	7	7	7	7	7	7	7	7	9	7
20	0	1	3	3	4	5	5	5	5	5	7	7	6	7	7	7	7	7	7	7	7	7	7	7	8	9	9	9	9	9	9	9
21	0	1	3	3	5	4	5	5	5	5	7	5	7	7	7	7	7	7	7	7	7	6	7	7	9	7	7	7	7	7	9	7

(x) Normal Play (S^+).

	0	1	2	3	4	5	6	7	8	9	10	11	12	13	14	15	16	17	18	19	20	21	22	23	24	25	26	27	28	29	30	31
0	0	0	0	0	0	0	0	0	0	0	0	0	0	0	0	0	0	0	0	0	0	0	0	0	0	0	0	0	0	0	0	0
1	0	1	2	2	1	2	2	1	2	2	1	2	2	1	2	2	1	2	2	1	2	2	1	2	2	1	2	2	1	2	2	1
2	0	2	3	4	4	4	3	4	4	4	2	3	4	4	4	3	4	4	4	2	3	4	4	4	3	4	4	4	2	3	4	4
3	0	2	4	3	4	4	4	4	3	4	4	4	4	2	3	4	4	4	4	2	4	3	4	4	4	4	3	4	4	4	4	2
4	0	1	4	4	5	6	6	6	6	6	3	4	4	3	4	6	6	6	6	6	4	4	5	6	6	6	6	6	3	4	4	3
5	0	2	4	4	6	5	6	6	6	6	4	6	5	6	6	6	6	5	6	6	6	6	6	6	4	6	4	5	6	6	6	6
6	0	2	3	4	6	6	5	6	6	6	6	5	6	6	6	6	5	6	6	6	6	5	6	6	6	6	5	6	6	6	6	4
7	0	1	4	4	6	6	6	5	6	6	6	6	6	3	4	6	6	6	5	6	6	6	6	5	6	6	6	6	4	5	6	6
8	0	2	4	3	6	6	6	6	7	8	8	8	8	8	8	8	8	8	6	7	8	8	8	8	8	8	8	8	8	6	5	6
9	0	2	4	4	6	6	6	6	8	7	8	8	8	8	8	8	8	8	6	8	5	6	6	6	6	7	8	8	8	8	8	8
10	0	1	2	4	3	4	6	6	8	8	5	6	6	6	6	6	6	7	8	8	8	8	8	8	8	8	6	7	8	8	8	8
11	0	2	3	4	4	6	5	6	8	8	6	6	6	6	5	6	6	8	6	8	6	6	7	8	8	8	8	8	8	8	8	8
12	0	2	4	4	4	5	6	6	8	8	6	6	7	8	8	8	8	8	8	8	8	7	8	8	8	8	8	8	8	8	6	7
13	0	1	4	2	3	6	6	3	8	8	6	6	8	7	8	8	8	8	8	8	8	8	8	7	8	8	8	8	8	8	8	8
14	0	2	4	3	4	6	6	4	8	8	6	5	8	8	6	6	6	8	7	8	8	8	8	8	8	8	8	8	7	8	8	8
15	0	2	3	4	6	6	6	6	8	8	6	6	8	8	6	7	8	8	8	8	8	8	8	8	7	8	8	8	8	8	8	8
16	0	1	4	4	6	6	5	6	8	8	6	6	8	8	6	8	7	8	8	8	8	8	8	8	8	8	7	8	8	8	8	8
17	0	2	4	4	6	5	6	6	8	8	7	8	8	8	8	8	8	8	8	8	7	8	8	8	8	8	8	8	8	7	8	8
18	0	2	4	4	6	6	6	5	6	6	8	6	8	8	7	8	8	8	8	8	8	8	8	8	8	9	A	A	A	A	A	A
19	0	1	2	2	6	6	6	6	7	8	8	8	8	8	8	8	8	8	8	9	A	A	A	A	A	A	A	A	A	A	A	A
20	0	2	3	4	4	6	6	6	8	5	8	6	8	8	8	8	8	7	8	A	8	8	8	8	8	A	8	8	8	8	7	8
21	0	2	4	3	4	6	5	6	8	6	8	6	7	8	8	8	8	8	8	A	8	7	8	8	8	A	8	8	8	8	8	8

(y) Misère Play (S^-). (A=10.)

Tables 4(x) and 4(y). Eating Cakes Impartially.

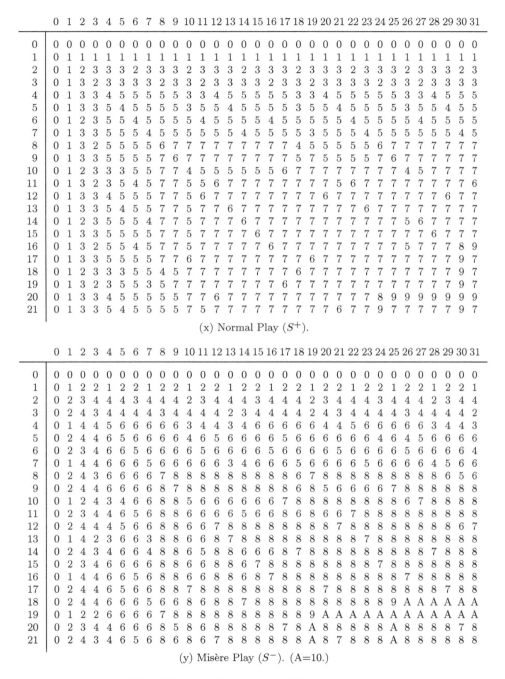

Extras

All the King's Horses on a Quarter-Infinite Board

Table 1 gave remotenesses of horses on an ordinary chessboard. Some of the values near the lower and right edges of the board are affected by the horse's inability to jump off the board. Table 5 shows the values when there are no lower and right edges; the patterns indicated by the lines continue indefinitely both downwards and to the right. That in Table 5(d) is hardest to see; the rows and columns have ultimate period 4 and saltus 4, but the character of the first, second and every third is different from that of the others.

You might like to work out corresponding tables for the suspense numbers. It's most sensible to allow a player to *pass* for a given horse when he can't move it, but his opponent can.

Cutting Your Cakes and Eating Them

Tables 6(a) and 6(b) give the normal and misère suspense numbers for the partizan game of Cutcakes, in which the rule is to cut every cake you can, so the game only ends when the cake is completely cut up. This implies that you make a pass move for a cake when you can't cut it, but your opponent can. Tables 6(x) and 6(y) are for the impartial version in which each player may cut in either direction.

Suspense numbers for Eatcake were given at the end of the chapter. Table 7 gives the corresponding remotenesses which are appropriate for the versions in which the game ends when the *first* strip (1 by n or n by 1 cake) is eaten. Table 7(y) for R^- (misère impartial) is not shown; you can find it by adding 1 to all the entries in Table 7(x), except the one for a 1 by 1 cake. That is $R^-(1,1) = R^+(1,1) = 1$; otherwise $R^- = R^+ + 1$.

(a) $R_L^+ R_R^+$ (normal play)

00	00	10	10	→					
00	00	10	10	→					
01	01	11	12	12	→				
01	01	21	22	22	32	32	→		
↓	↓	21	22	22	32	32	→		
		↓	23	23	33	34	34	→	
			23	23	43	44	44	54	54
			↓	↓	43	44	44	54	54
					↓	45	45	55	56
						45	45	65	66

(b) $R_L^- R_R^-$ (misère play)

00	00	10	10	→					
00	00	20	30	30	→				
01	02	22	42	52	52	→			
01	03	24	44	64	74	74	→		
↓	03	25	46	66	86	96	96	→	
	↓	25	47	68	88	A8	B8	B8	→
		↓	47	69	8A	AA	CA	DA	DA
			↓	69	8B	AC	CC	EC	FC
				↓	8B	AD	CE	EE	GE
					↓	AD	CF	EG	GG

(c) $R_L^+ R_R^+$ (normal with pass)

00	00	12	12	34		56		78		9A	
00	00	12	12								
21	21	11	14	34	36	56	58	78	7A	9A	9C
21	21	41	44		56		78		9A		
43		43									
		63	65		55	58	78	7A	9A	9C	
65		65			85	88		9A		BC	
		85	87		87						
87		87			A7	A9		99	9C	BC	
		A7	A9		A9			C9	CC		
A9		A9			C9	CB		CB			
		C9									

(d) $R_L^- R_R^-$ (misère with pass)

00	00	12	12	45	56	56	56	89	9A	9A	9A
00	00	23	34	34	34	67	78	78	78	AB	BC
21	32	22	42	42	56	56	56	86	9A	9A	9A
21	43	24	44	54	64	74	78	78	78	78	BC
54	43	24	45	66	76	86	76	96	9A	9A	9A
65	43	65	46	67	66	98	A8	98	B8	98	B8
65	76	65	47	68	89	88	BA	C8	BA	DA	BA
65	87	65	87	67	8A	AB	AA	DC	EA	DC	FA
98	87	68	87	69	89	8C	CD	CC	FE	GC	FE
A9	87	A9	87	A9	8B	AB	AE	EF	EE	HG	IE
A9	BA	A9	87	A9	89	AD	CD	CG	GH	GG	JI
A9	CB	A9	CB	A9	8B	AB	AF	EF	EI	IJ	II

(x) R^+ (normal impartial)

0	0	1	1	2	2	3	3	4	4	5	5
0	0	1	1	2	2	3	3	4	4	5	5
1	1	1	1	3	3	3	3	5	5	5	5
1	1	1	3	3	3	3	5	5	5	5	7
2	2	3	3	4	4	5	5	6	6	7	7
2	2	3	3	4	4	5	5	6	6	7	7
3	3	3	3	5	5	5	5	7	7	7	7
3	3	3	5	5	5	5	7	7	7	7	9
4	4	5	5	6	6	7	7	8	8	9	9
4	4	5	5	6	6	7	7	8	8	9	9
5	5	5	5	7	7	7	7	9	9	9	9

(y) R^- (misère impartial)

0	0	1	1	3	4	3	3	5	6	5	5
0	0	2	3	2	2	4	5	4	4	6	7
1	2	2	2	2	4	4	4	4	6	6	6
1	3	2	3	4	5	6	5	6	7	8	7
3	2	2	4	5	4	5	6	7	6	7	8
4	2	4	5	4	7	6	7	6	9	8	9
3	4	4	6	5	6	7	6	7	8	9	8
3	5	4	5	6	7	6	7	8	9	8	9
5	4	4	6	7	6	7	8	9	8	9	A
6	4	6	7	6	9	8	9	8	9	A	B
5	6	6	8	7	8	9	8	9	A	B	A

Table 5. All the King's Horses H=17, I=18, J=19.)

Cut every cake you can (i.e. "with pass"); the game ends when the cake is all in 1×1 bits. (a) last cut wins, (b) last cut loses.

(a) $S_L^+ S_R^+$

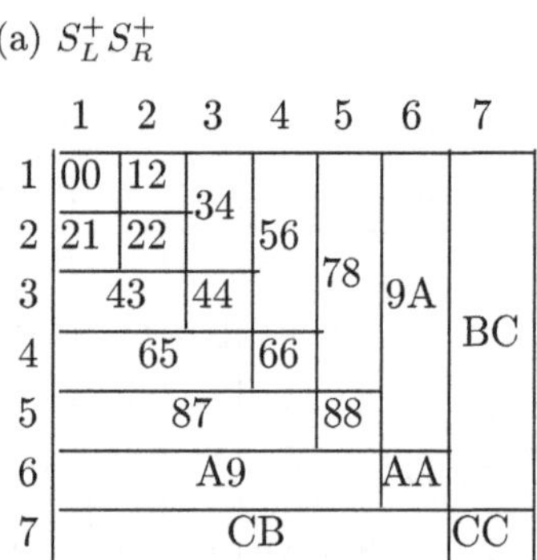

	1	2	3	4	5	6	7
1	00	12	34				
2	21	22		56			
3	43		44		78	9A	
4	65			66			BC
5	87				88		
6	A9					AA	
7	CB						CC

(b) $S_L^- S_R^-$

	1	2	3	4	5	6	7	8	9	10	11	12	13
1	00	12	34	56	78	9A	BC	DE	FG	HI	JK	LM	NO
2	21	22	34	36									
3	43	43	44	46	58	5A	7C	7E					
4	65	63	64	66					9G	9I	BK	BM	
5	87		85		88	6A	6C	6E					DO
6	A9		A5		A6	AA	6C	6E	7G	7I			
7	CB		C7		C6	C6	CC	8E	8G	8I	7K	7M	
8	ED		E7		E6	E6	E8	EE	8G	8I	7K	7M	
9	GF		G9			G7	G8	G8	GG	AI	AK		
10	IH		I9			I7	I8	I8	IA	II			
11	KJ			KB			K7	K7	KA				
12	ML			MB			M7	M7	M8				
13	ON				OD			O9	OA				
14	QP				QD			Q9	Q9				
15	SR				SF				S9				
16	UT				UF				U9				

(x) S^+ (normal impartial)

	1	2	3	4	5	6	7	8	9	10	11
1	0	1	2	3	3	3	4	5	5	5	5
2	1	2	3	3	4	5	5	5	5	5	6
3	2	3	4	5	5	5	6	7	7	7	7
4	3	3	5	4	5	5	7	5	6	7	7
5	3	4	5	5	6	7	7	7	7	7	8
6	3	5	5	5	7	6	7	7	7	7	9
7	4	5	6	7	7	7	8	9	9	9	9
8	5	5	7	5	7	7	9	6	7	7	9
9	5	5	7	6	7	7	9	7	8	9	9
10	5	5	7	7	7	7	9	7	9	8	9
11	5	6	7	7	8	9	9	9	9	9	A

(y) S^- (misère impartial)

	1	2	3	4	5	6	7	8	9	10	11	12
1	0	1	2	2	3	4	4	4	4	4	5	6
2	1	2	3	4	4	4	5	6	6	6	6	6
3	2	3	4	4	5	6	6	6	6	6	7	8
4	2	4	4	5	6	6	6	6	7	8	8	8
5	3	4	5	6	6	6	7	8	8	8	8	8
6	4	4	6	6	6	7	8	8	8	8	8	8
7	4	5	6	6	7	8	8	8	8	8	9	A
8	4	6	6	6	8	8	8	7	8	8	A	8
9	4	6	6	7	8	8	8	8	9	A	A	A
10	4	6	6	8	8	8	8	8	A	9	A	A
11	5	6	7	8	8	8	9	A	A	A	A	A

(The boxes indicate curves formed by the "most NW" $\mathcal{P}$-positions.)

Table 6. Suspense Numbers for Normal, Misère and Impartial Cutcakes. (A=10, B=11, C=12, ...)

(a) $R_L^+ R_R^+$ (first eaten strip wins)

	0	1	2	3	4	5	6
0	∞	∞	∞	∞	∞	∞	∞
1	∞	11	21	→			
2	∞	12					
3	∞	↓			33		
4	∞						
5	∞						
6	∞						

(b) $R_L^- R_R^-$ (first eaten strip loses)

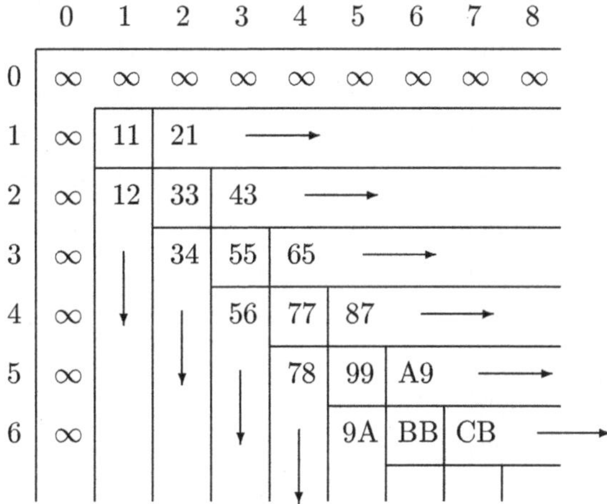

	0	1	2	3	4	5	6	7	8
0	∞	∞	∞	∞	∞	∞	∞	∞	∞
1	∞	11	21	→					
2	∞	12	33	43	→				
3	∞	↓	34	55	65	→			
4	∞		↓	56	77	87	→		
5	∞			↓	78	99	A9	→	
6	∞				↓	9A	BB	CB	→

Tables 7(a) and (b). Partizan Remotenesses for Eatcakes.

(x) R^+ (normal impartial)

	0	1	2	3	4	5	6	7	8	9	10	11	12	13	14	15	16
0	∞	∞	∞	∞	∞	∞		→									
1	∞	1	1	1	1	1	1		→								
2	∞	1	2	3	4	3	3	3		→							
3	∞	1	3	4	5	6	7	5	5	5		→					
4	∞	1	4	5	6	5	5	5	5	5	5		→				
5	∞	1	3	6	5	6	7	7	8	7	7	7	7	7		→	
6	↓	1	3	7	5	7	8	9	9	10	11	12	13	9	9	9	9
7		↓	3	5	5	7	9	10	11	11	12	13	14	15	16	11	11
8			↓	5	5	8	9	11	12	13	13	9	9	9	9	9	9
9				5	5	7	10	11	13	14	15	16	17	11	11	11	11
10				↓	5	7	11	12	13	15	16	17	18	19	20	13	13
11					↓	7	12	13	9	16	17	18	19	13	13	14	13
12						7	13	14	9	17	18	19	20	21	22	15	16
13						7	9	15	9	11	19	13	21	22	23	24	17
14						↓	9	16	9	11	20	13	22	23	24	17	18
15							9	11	9	11	13	14	15	24	17	18	19
16							↓	11	9	11	13	13	16	17	18	19	20
17								11	9	11	13	13	15	18	19	20	21
18								↓	9	11	13	13	15	19	17	21	22
19									↓	11	13	13	15	20	21	22	23
20										↓	13	13	15	21	17	23	24
21											↓	13	15	22	17	24	25
22												↓	15	23	17	25	26
23													↓	24	17	15	27
24														25	17	15	28
25														26	↓	15	17
26														27		15	17
27														23		↓	17
28														19			17
29														23			↓
30														19			
31														19			
32														19			
33														19			

Table 7(x). Impartial Remotenesses for Eatcakes.

17	18	19	20	21	22	23	24	25	26	27	28	29	30	31	32	33	34	
	→																	
11		→																
9	9		→															
11	11	11		→														
13	13	13	13		→													
13	13	13	13	13		→												
15	15	15	15	15	15		→											
18	19	20	21	22	23	24	25	26	27	23	19	23	19	19	19	19	21	
19	17	21	17	17	17	17	17		→									
20	21	22	23	24	25	15	15	15	15		→							
21	22	23	24	25	26	27	28	17	17	17	17		→					
22	23	24	25	26	27	28	29	30	31	19	19	19	19		→			
23	24	25	26	27	28	29	30	31	32	33	34	35	36	37	38	23		→
24	25	26	27	28	29	30	31	32	33	21	21		→					
25	26	27	28	29	30	31	32	33	34	35	36	37	38	39	40	25		→
26	27	28	29	30	31	32	33	34	35	23	37	23	23		→			
27	28	29	30	31	32	33	34	35	36	37	38	39	40	41	42	27		→
28	29	30	31	32	33	34	35	36	37	25	39	40	25	25		→		
29	30	31	32	33	34	35	36	37	38	39	40	41	42	43	44	29		→
30	31	32	33	34	35	36	37	38	39	27	41	42	43	27	27		→	
31	32	33	34	35	36	37	38	39	40	41	42	43	44	45	46	47	48	
19	33	21	35	23	37	25	39	27	41	42	43	44	45	46	47	48	49	
19	34	21	36	37	38	39	40	41	42	43	44	45	46	47	48	49	50	
19	35	↓	37	23	39	40	41	42	43	44	45	46	47	48	49	50	51	
19	36		38	23	40	25	42	43	44	45	46	47	48	49	50	51	52	
↓	37		39	↓	41	25	43	27	45	46	47	48	49	50	51	52	53	
	38		40		42	↓	44	27	46	47	48	49	50	51	52	53	54	
	23		25		27		29	↓	47	48	49	50	51	52	53	54	55	

References and Further Reading

J. H. Conway, *On Numbers and Games*, Second Edition, A K Peters, Ltd., 2001, Chapter 14.

Martin Gardner, Mathematical Games: cram, crosscram and quadraphage: New games having elusive winning strategies, *Sci. Amer.*, **230** #2(Feb. 1974) 106–108.

C. A. B. Smith, Graphs and composite games, *J. Combin. Theory*, **1**(1966) 51–81; *MR* **33** #2572.

H. Steinhaus, Definicje potrzebne do teorji gry i pościgu, *Myśl. Akad. Lwów*, **1**#1(1925) 13–14; reprinted as Definitions for a theory of games and pursuit, *Naval Res. Logist. Quart.*, **7**(1960) 105–108.

-10-

Hot Battles Followed by Cold Wars

And through the heat of conflict, keeps the law
In calmness made and sees what he foresaw.
William Wordsworth, *Character of the Happy Warrior.*

When the rules of a compound game allow you to move in any desired number of component subgames, we get what has been called a **selective compound**; we shall call it the **union**. If there are any *hot* components about, both players will naturally move in all of them, and they would like this part of the game to last as long as possible. When all components are cold, they will move only in the least disadvantageous one. So a union of games, if well played, will consist of the slow join of its hot parts, followed by the ordinary sum of the residual cold games.

Hotcakes

Figure 1. Rita Taking Her Turn.

Mother has been making cakes again and this time they are quite hot. In Cutcake (Chapter 2), Left made a vertical, or Rita a horizontal, cut in just one cake. In Cutcakes (Chapter 9), they made such cuts in all the cakes they could. Now they may cut whichever cakes they like, but to make it more interesting, after each cake is cut, just one of its two parts is turned through a right angle before being put back on the table. In Fig. 1 Rita has just made a horizontal cut and is seen turning one of the resulting two pieces of cake.

Unions of Games

Hotcakes is the selective compound or union of a number of component one-cake games. More generally there is such a compound

$$G \vee H \vee K \vee \ldots \qquad (\text{“}G \text{ or } H \text{ or } K \text{ or } \ldots\text{”})$$

of any number of component games G, H, K, $\ldots$. When it is his turn to move, a player **selects** some of the components (at least one, maybe all) and then makes moves, legal for him, in each game of his selection. When a player cannot move, the game ends and that player is the loser according to the normal play rule. The selective theory (of unions) is a mixture of the disjunctive theory (of sums) and the conjunctive theory (of slow joins).

Cold Games – Numbers Are Still Numbers

Some games behave exactly like ordinary disjunctive sums. In Coldcakes we allow Lefty or Rita to cut as many cakes as they like, but we remove the new turning requirement. How does this differ from Cutcake, which they played in Chapter 2?

Not at all! Since every value in that game was a number, they were never anxious to make moves and will now be even less keen to make several moves at once. Each player will move in only one component—that which does him least harm—and the union becomes an ordinary sum. The same will happen in any game (such as Blue-Red Hackenbush) in which all the values are numbers.

Hot Games – The Battle Is Joined!

In a union of games there may be several hot components, in which the players want to move, and some cold ones, in which they don't. So they won't touch any cold component but will move in *all* the hot components while any such remaain. We call this first part of the union the **hot battle**. Since the players make all the hot moves they can, it is a *join* of smaller hot battles for the hot components, and since it only ends when the *last* of these does, it is a *slow* join.

After the hot battle we have the **cold war**, which is just an ordinary addition sum, since all the components are numbers and the players will only move in one at a time.

Tolls, Timers and Tallies

Table 1 shows the left and right tallies for Hotcakes. For the 2 by 3 cake the entry is

$$1_2 0_1.$$

The **left tally**, 1_2, consists of a **toll**, 1, and a **timer**, 2, and means that if Left starts, we shall reach a cold war position of 1 after a hot battle of 2 moves. If instead Right starts we reach a cold war position of 0 after a battle of 1 move. These two little battles are illustrated in Fig. 2. It is easy to tell when the cold war begins because collections of Hotcakes only get cold when all the cakes are strips. The entry for a strip in Table 1 is therefore a single number, x, which should be regarded as an abbreviation for the pair of tallies $x_0 x_0$.

What is the fate of Fig. 3?

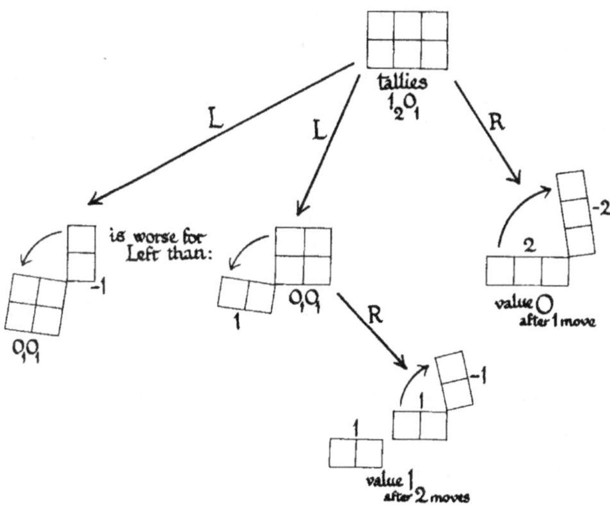

Figure 2. Battles for a Small Hotcake.

Here the left tallies are

$$4_3, \qquad 1_2, \qquad 0_5,$$

and so if Left moves first the value when the cold war starts will be $4+1+0=5$. And since the longest component battle (and therefore the whole battle) will last for 5 moves, the left

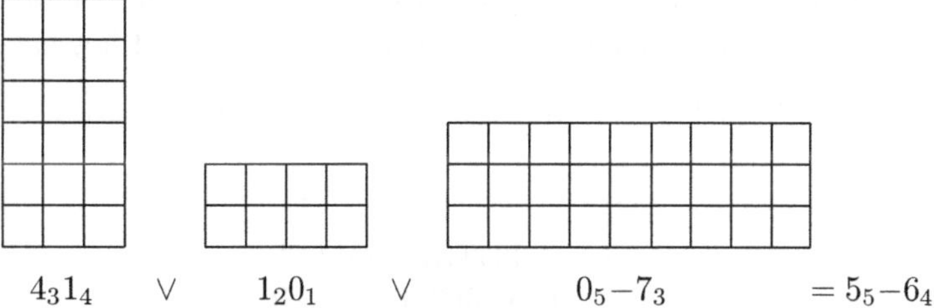

$$4_3 1_4 \quad \vee \quad 1_2 0_1 \quad \vee \quad 0_5{-}7_3 \qquad = 5_5{-}6_4.$$

Figure 3. A Hotcake Position.

	1	2	3	4	5	6	7	8	9
1	0	1	2	3	4	5	6	7	8
2	-1	$0_{1}0_{1}$	$1_{2}0_{1}$	$1_{2}0_{1}$	$1_{2}0_{1}$	$1_{2}0_{1}$	$1_{2}0_{1}$	$1_{2}0_{1}$	$1_{2}0_{1}$
3	-2	$0_{1}-1_{2}$	$1_{3}-1_{3}$	$1_{4}-2_{3}$	$0_{4}-3_{3}$	$-1_{4}-4_{3}$	$-1_{5}-5_{3}$	$0_{5}-6_{3}$	$0_{5}-7_{3}$
4	-3	$0_{1}-1_{2}$	$2_{3}-1_{4}$	$2_{5}-2_{5}$	$1_{6}-4_{5}$	$-1_{6}-6_{5}$	$-2_{7}-7_{6}$	$0_{7}-7_{6}$	$1_{8}-8_{6}$
5	-4	$0_{1}-1_{2}$	$3_{3}0_{4}$	$4_{5}-1_{6}$	$3_{7}-3_{7}$	$1_{8}-6_{7}$	$-2_{8}-8_{8}$	$-2_{9}-7_{8}$	$1_{9}-7_{9}$
6	-5	$0_{1}-1_{2}$	$4_{3}1_{4}$	$6_{5}1_{6}$	$6_{7}-1_{8}$	$4_{9}-4_{9}$	$1_{10}-8_{9}$	$-3_{10}-9_{10}$	$-2_{11}-7_{10}$
7	-6	$0_{1}-1_{2}$	$5_{3}1_{5}$	$7_{6}2_{7}$	$8_{8}2_{8}$	$8_{9}-1_{10}$	$5_{11}-5_{11}$	$1_{12}-10_{11}$	$-4_{12}-10_{12}$
8	-7	$0_{1}-1_{2}$	$6_{3}0_{5}$	$7_{6}0_{7}$	$7_{8}2_{9}$	$9_{10}3_{10}$	$10_{11}-1_{12}$	$6_{13}-6_{13}$	$1_{14}-12_{13}$
9	-8	$0_{1}-1_{2}$	$7_{3}0_{5}$	$8_{6}-1_{8}$	$7_{9}-1_{9}$	$7_{10}2_{11}$	$10_{12}4_{12}$	$12_{13}-1_{14}$	$7_{15}-7_{15}$
10	-9	$0_{1}-1_{2}$	$8_{3}0_{6}$	$9_{7}-1_{8}$	$8_{9}-2_{9}$	$7_{10}-2_{11}$	$7_{12}2_{13}$	$11_{14}4_{11}$	$13_{12}-1_{16}$
11	-10	$0_{1}-1_{2}$	$9_{3}1_{6}$	$11_{7}0_{8}$	$10_{9}-2_{10}$	$8_{11}-4_{11}$	$6_{12}-3_{13}$	$7_{14}2_{15}$	$12_{16}4_{13}$
12	-11	$0_{1}-1_{2}$	$10_{3}1_{7}$	$12_{8}2_{9}$	$13_{10}-1_{10}$	$10_{11}-4_{11}$	$7_{12}-5_{13}$	$6_{14}-3_{12}$	$8_{13}2_{17}$
13	-12	$0_{1}-1_{2}$	$11_{3}0_{7}$	$12_{8}1_{9}$	$13_{10}1_{10}$	$13_{11}-3_{12}$	$9_{13}-6_{13}$	$6_{14}-4_{12}$	$8_{13}-3_{14}$
14	-13	$0_{1}-1_{2}$	$12_{3}0_{7}$	$13_{8}-1_{10}$	$12_{11}2_{11}$	$15_{12}-1_{12}$	$12_{13}-6_{13}$	$7_{14}-4_{12}$	$9_{13}-4_{14}$
15	-14	$0_{1}-1_{2}$	$13_{3}0_{8}$	$14_{9}-1_{10}$	$13_{11}0_{11}$	$14_{12}2_{12}$	$16_{13}-4_{14}$	$10_{15}-6_{12}$	$8_{13}-3_{14}$
16	-15	$0_{1}-1_{2}$	$14_{3}1_{8}$	$16_{9}0_{10}$	$15_{11}-1_{11}$	$14_{12}2_{13}$	$17_{14}-1_{14}$	$14_{15}-7_{12}$	$8_{13}-3_{14}$
17	-16	$0_{1}-1_{2}$	$15_{3}1_{9}$	$17_{10}1_{10}$	$17_{11}-2_{12}$	$14_{13}-1_{13}$	$15_{14}3_{14}$	$19_{15}-5_{16}$	$11_{17}-6_{14}$
18	-17	$0_{1}-1_{2}$	$16_{3}0_{9}$	$17_{10}1_{11}$	$18_{12}-1_{12}$	$16_{13}-3_{13}$	$14_{14}2_{15}$	$19_{16}-1_{16}$	$16_{17}-8_{14}$
19	-18	$0_{1}-1_{2}$	$17_{3}0_{9}$	$18_{10}0_{11}$	$18_{12}1_{12}$	$19_{13}-3_{13}$	$15_{14}-2_{15}$	$16_{16}1_{13}$	$19_{14}-6_{18}$
20	-19	$0_{1}-1_{2}$	$18_{3}0_{10}$	$19_{11}0_{11}$	$19_{13}2_{13}$	$21_{14}-3_{14}$	$16_{15}-4_{15}$	$15_{16}1_{13}$	$20_{14}-1_{15}$
21	-20	$0_{1}-1_{2}$	$19_{3}1_{10}$	$21_{11}0_{12}$	$20_{13}0_{13}$	$20_{14}-1_{14}$	$19_{15}-5_{15}$	$15_{16}0_{14}$	$20_{15}0_{15}$
22	-21	$0_{1}-1_{2}$	$20_{3}1_{11}$	$22_{12}1_{12}$	$22_{13}-1_{13}$	$20_{14}2_{14}$	$23_{15}-5_{15}$	$16_{16}-1_{14}$	$20_{15}-1_{15}$
23	-22	$0_{1}-1_{2}$	$21_{3}0_{11}$	$22_{12}1_{13}$	$23_{14}-2_{14}$	$20_{15}2_{15}$	$24_{16}-4_{16}$	$18_{17}-1_{17}$	$21_{18}-1_{15}$
24	-23	$0_{1}-1_{2}$	$22_{3}0_{11}$	$23_{12}0_{13}$	$23_{14}-1_{14}$	$22_{15}-1_{15}$	$22_{16}-1_{16}$	$22_{17}-3_{14}$	$20_{15}0_{16}$
25	-24	$0_{1}-1_{2}$	$23_{3}0_{12}$	$24_{13}0_{13}$	$24_{14}1_{14}$	$25_{15}-3_{15}$	$21_{16}2_{16}$	$26_{17}-4_{14}$	$20_{15}1_{16}$

Table 1. Left and Right Tallies for Hotcakes.

tally of the position is 5_5. The right tally is

$$(1+0-7)_4 = -6_4,$$

since the component right tallies are

$$1_4, \qquad 0_1, \qquad -7_3,$$

and the battle this time lasts for 4 moves.

For a union of positions with tallies

$$u_a x_i, \qquad v_b y_j, \qquad w_c z_k, \qquad \ldots$$

we find both left and right tallies by *adding* the tolls (which refer to an eventual *sum* of cold wars) and taking the *largest* timers (which refer to a *slow join* of hot battles), obtaining

$$(u+v+w+\cdots)_{\max(a,b,c,\ldots)}(x+y+z+\cdots)_{\max(i,j,k,\ldots)}$$

TOTAL TOLLS!
TOPMOST TIMERS!

Which Is the Best Option?

Since after Left's move it will be Right's turn, Left need only consider the *right* tallies of his options and make a shortlist of those with the largest toll. Right's shortlist consists of those *left* tallies of his options that have least toll. Both players should now choose options from their shortlists by the normal suspense rule of Chapter 9—take the largest even timer if there is one and otherwise the smallest odd timer.

> Shortlist all
> G^L with GREATEST RIGHT TOLL
> and all
> G^R with LEAST LEFT TOLL
> and, from the corresponding
> timers on each side, take the
> LARGEST EVEN else SMALLEST ODD

CHOOSING BEST OPTIONS

All that remains to be explained is how to find the tallies of an arbitrary position from those of its options. Before we do this we will use what we already know to find the best options for the 9 by 3 cake. According to the rules of Hotcakes this has options as shown in

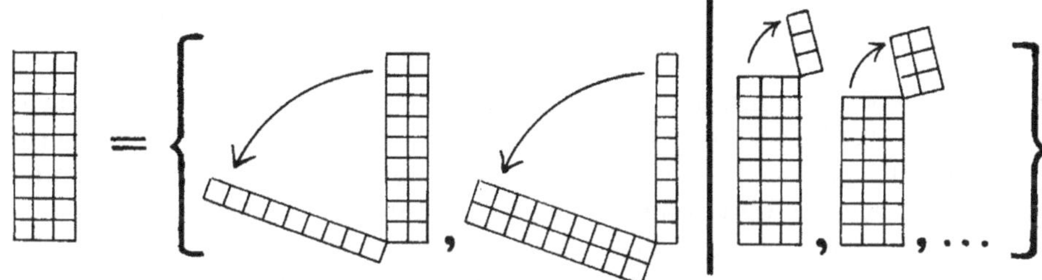

or, in symbols,

$$9 \text{ by } 3 = \left\{ \begin{array}{c|c} & 8 \text{ by } 3 \vee 3 \text{ by } 1 \\ & 7 \text{ by } 3 \vee 3 \text{ by } 2 \\ & 6 \text{ by } 3 \vee 3 \text{ by } 3 \\ 1 \text{ by } 9 \vee 9 \text{ by } 2 & 5 \text{ by } 3 \vee 3 \text{ by } 4 \\ 2 \text{ by } 9 \vee 9 \text{ by } 1 & 4 \text{ by } 3 \vee 3 \text{ by } 5 \\ & 3 \text{ by } 3 \vee 3 \text{ by } 6 \\ & 2 \text{ by } 3 \vee 3 \text{ by } 7 \\ & 1 \text{ by } 3 \vee 3 \text{ by } 8 \end{array} \right\},$$

with tallies

$$\left\{\begin{array}{r} \\ \\ \Downarrow \\ 8 \;\vee 0_1{-}1_2 = \;\; 8_1\mathbf{7_2} \\ 1_20_1 \vee \;\; -8 \;\; = -7_2{-}\mathbf{8_1} \\ \\ \\ \\ \end{array}\right|\begin{array}{rcl} 6_30_5 \vee & -2 & = \mathbf{4_3}{-}2_5 \\ 5_31_5 \vee & 0_1{-}1_2 & = \mathbf{5_3}0_5 \\ 4_31_4 \vee & 1_3{-}1_3 & = \mathbf{5_3}0_4 \\ 3_30_4 \vee & 1_4{-}2_2 & = \mathbf{4_4}{-}2_4 \\ 2_3{-}1_4 \vee & 0_4{-}3_3 & = \mathbf{2_4}{-}4_4 \\ 1_3{-}1_3 \vee & -1_4{-}4_3 & = \mathbf{0_4}{-}5_3 \\ 1_20_1 \vee & -1_5{-}5_3 & = \mathbf{0_5}{-}5_3 \\ 2 \vee & 0_5{-}6_3 & = \mathbf{2_5}{-}4_3 \end{array}\left.\begin{array}{l} \\ \\ \\ \\ \\ \Leftarrow \\ \\ \\ \end{array}\right\}$$

Of course, because Left only refers to right tallies and Right only to left ones, the bold-face figures are all we need. From them Left shortlists only 7_2 and Right only 0_4 and 0_5 from which she chooses 0_4. Why is this? If Right made a move to 0_5 the battle would have 5 more moves,

Left, Right, Left, Right, Left,

and Right would be forced to move first in the cold war. She therefore prefers the move to 0_4 with 4 battle moves,

Left, Right, Left, Right,

after which Left must make the first cold war move. We conclude that the players' best options have the tallies shown in

$$\{..7_2|0_4..\}.$$

in which the pairs of dots represent the (irrelevant) Left tally of the best Left option, and Right tally of the best Right option.

Hot Positions

When we know the best options, how do we find the new tallies? In the example there is no difficulty. Since $7 > 0$, this position is still hot, so the battle is not yet over. Indeed we can see just how long it will last: If Left makes the first move there will be just 2 more battle moves, while a first move by Right would be followed by 4 further battle moves. The hot battle therefore lasts 3 or 5 moves in the two cases and the tallies are 7_30_5.

The same argument works for hot positions in any game when the best options have been found:

> For a position
> $\{..x_a|y_b..\}$ with $x > y$,
> the tallies are
> $x_{a+1}y_{b+1}$

TALLY RULE FOR HOT POSITIONS

Cold Positions

But if $x < y$ the position is cold and must be a number. Which number is it? We find out by a version of the Simplicity Rule.

We shall discuss the cases in which Left's option has right tally $\frac{1}{2}_7$ or $\frac{1}{2}_8$ and Right's has left tally 1_3 or 1_4.

Since 7 is odd and 8 is even,

$\frac{1}{2}_8$ is *better for Left* than $\frac{1}{2}$ (i.e. $\frac{1}{2}_8 > \frac{1}{2}$),

while

$\frac{1}{2}_7$ is *worse for Left* than $\frac{1}{2}$ (i.e. $\frac{1}{2}_7 < \frac{1}{2}$),

and for similar reasons

1_4 is *better for Right* than 1 (i.e. $1_4 < 1$),

while

1_3 is *worse for Right* than 1 (i.e. $1_3 > 1$),

[Recall that "greater than" is "better for Left", "worse for Right".] We can now see from the picture

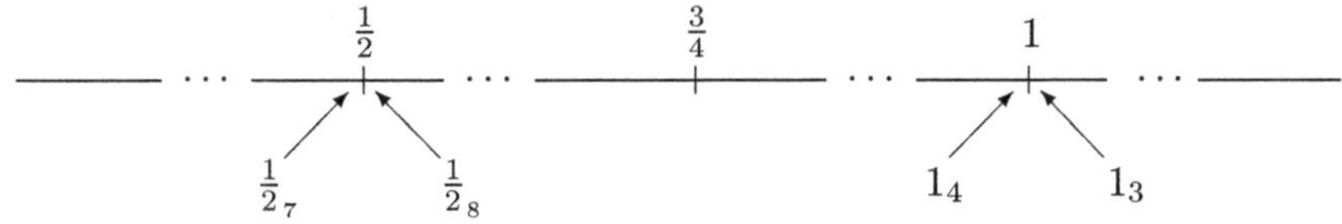

that the simplest numbers in the appropriate ranges are

$$\begin{array}{rl} \frac{3}{4} & \text{for} \quad \{..\tfrac{1}{2}_8 | 1_4 ..\}, \\ 1 & \text{for} \quad \{..\tfrac{1}{2}_7 | 1_3 ..\}, \\ 1 & \text{for} \quad \{..\tfrac{1}{2}_8 | 1_3 ..\}, \\ \text{and } \frac{1}{2} & \text{for} \quad \{..\tfrac{1}{2}_7 | 1_4 ..\}. \end{array}$$

For cold positions in general:

> For a position
> $\{..x_a \mid y_b ..\}$ with $x < y$,
> the tallies are $z_0 z_0$, where z is the
> simplest number in the range
> $x < z < y$ if a and b are both even,
> $x \le z \le y$ if a and b are both odd,
> $x < z \le y$ if a is even and b is odd,
> $x \le z < y$ if a is odd and b is even.

TALLY RULES FOR COLD POSITIONS

In short:

> *odd* timers *admit* their tolls,
> *even* timers *exclude* their tolls,
> as candidates for the simplest number.

Or, shorter still:

ODD ADMITS!
EVEN EVICTS!

Of course such positions, being cold, are really numbers, and a number z has tallies $z_0 z_0$ because it is already *in* the cold war.

Tepid Positions

When Left's and Right's options both have the same toll, the rules are more delicate:

For a position
$\{..x_a|x_b..\}$
with equal tolls, the tallies are
$x_{a+1}x_{b+1}$ if a and b are both even,
$x_0 x_0$ if a and b are both odd,
$x_{\max(a+1,b+2)}x_{b+1}$ if a is even and b is odd,
$x_{a+1}x_{\max((b+1,a+2)}$ if a is odd and b is even.

TALLY RULES FOR TEPID POSITIONS

It is easier in practice to ask first whether the tentative tallies

$$x_{a+1}y_{b+1}$$

obey the *Lukewarmth Commandment*:

IF TOLLS BE EQUAL,
THOU SHALT NOT PERMIT AN
EVEN TIMER UNLESS THERE
BE A GREATER *ODD* ONE

Then tallies disobeying the commandment are of two types and should be corrected as follows:

If the timers $a+1$, $b+1$ are both even, the correct value is x with tallies $x_0 x_0$.
If one of $a+1$, $b+1$ is an even number, e, and the other is a smaller odd number, this odd number should be increased to $e+1$.

These rather puzzling rules must be explained in terms of battles only, since we know the value x at which the cold war will start. The important question is: who must start it? To decide when he himself wants to enter the fray, the erudite player will of course be guided by the maxim of Chapter 9:

> GREATER GOOD! (odd)
> LESSER EVIL! (even)

Let us see how this leads to the above rules by studying four simple examples.

<table>
<tr><td colspan="4">For the battle of the example</td></tr>
<tr><td>$\{..x_8|x_4..\}$</td><td>$\{..x_7|x_3..\}$</td><td>$\{..x_8|x_3..\}$</td><td>$\{..x_7|x_4..\}$</td></tr>
<tr><td colspan="4">Right sees, if she moves first, a</td></tr>
<tr><td>win</td><td>loss</td><td>short loss</td><td>short win</td></tr>
<tr><td colspan="4">as against the</td></tr>
<tr><td>loss</td><td>win</td><td>longer loss</td><td>longer win</td></tr>
<tr><td colspan="4">that happens if she leaves the opening move to Left. She therefore</td></tr>
<tr><td>does</td><td>does not</td><td>does</td><td>does not</td></tr>
<tr><td colspan="4">want to make the first move. Left, by moving first, can guarantee a</td></tr>
<tr><td>win</td><td>loss</td><td>long win</td><td>loss of definite length;</td></tr>
<tr><td colspan="4">as against the possible</td></tr>
<tr><td>loss</td><td>win</td><td>short win</td><td>indefinitely long loss</td></tr>
<tr><td colspan="4">if he leaves it to Right, who may or may not respond. He therefore</td></tr>
<tr><td>does</td><td>does not</td><td>does</td><td>does</td></tr>
<tr><td colspan="4">prefer to move first. The resulting tallies are</td></tr>
<tr><td>x_9x_5</td><td>x_0x_0</td><td>x_9x_4</td><td>x_8x_9</td></tr>
<tr><td colspan="4">because the game is</td></tr>
<tr><td>hot</td><td>cold</td><td>(fairly hot)</td><td>both hot and cold!</td></tr>
</table>

More detailed explanations:

The first example $\{..x_8|x_4..\}$ is *patently* hot since $x_8 > x > x_4$, so the final timers are got by adding 1 as usual.

The second example $\{..x_7|x_3..\}$ is just as patently cold, since $x_7 < x < x_3$ and we get the number x by the Simplicity Rule.

In the third example $\{..x_8|x_3..\}$ the heat is latent. Although Left definitely wins, both players want to move: Right to take the lesser evil and Left to ensure that his good is greater. The timers are still obtained by adding 1.

The last example $\{..x_7|x_4..\}$ is the most subtle! Right, who is sure to win the battle, has all the time in the world and would be delighted if nobody touched this game for a long time, since it might then be the last battle to end. Left, therefore, will move as soon as he can. If he is due to move, he moves to x_7 and the battle will last for 8 moves in all. But if it is Right's turn, and battles rage elsewhere, she should make a move without involving this component; and with Left then due to move there will be 8 more moves in this battle, making 9 in all. The tallies are therefore x_8x_9.

Tally Truths Totally Told

We can now give a complete summary of all our rules for working with tallies:

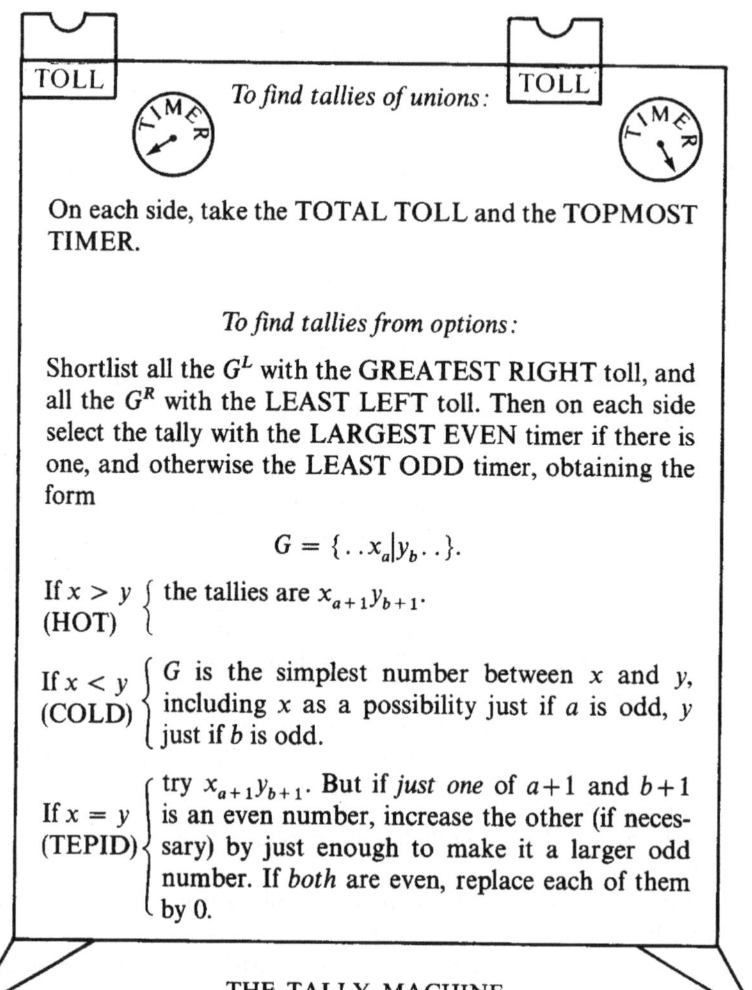

TOLL
TOLL
TIMER
TIMER
To find tallies of unions:
On each side, take the TOTAL TOLL and the TOPMOST TIMER.
To find tallies from options:
Shortlist all the G^L with the GREATEST RIGHT toll, and all the G^R with the LEAST LEFT toll. Then on each side select the tally with the LARGEST EVEN timer if there is one, and otherwise the LEAST ODD timer, obtaining the form
$G = \{..x_a|y_b..\}.$
If $x > y$ (HOT) the tallies are $x_{a+1}y_{b+1}$.
If $x < y$ (COLD) G is the simplest number between x and y, including x as a possibility just if a is odd, y just if b is odd.
If $x = y$ (TEPID) try $x_{a+1}y_{b+1}$. But if *just one* of $a+1$ and $b+1$ is an even number, increase the other (if necessary) by just enough to make it a larger odd number. If *both* are even, replace each of them by 0.

THE TALLY MACHINE

A Tepid Game

Coolcakes is another eating game. It is the selective version of Eatcakes, which we met in Chapter 9. Across each one of as many cakes as she likes, Rita eats away a horizontal strip. Lefty eats vertical strips in a similar way. Table 2 gives the tallies, and the displayed pattern can be shown to continue.

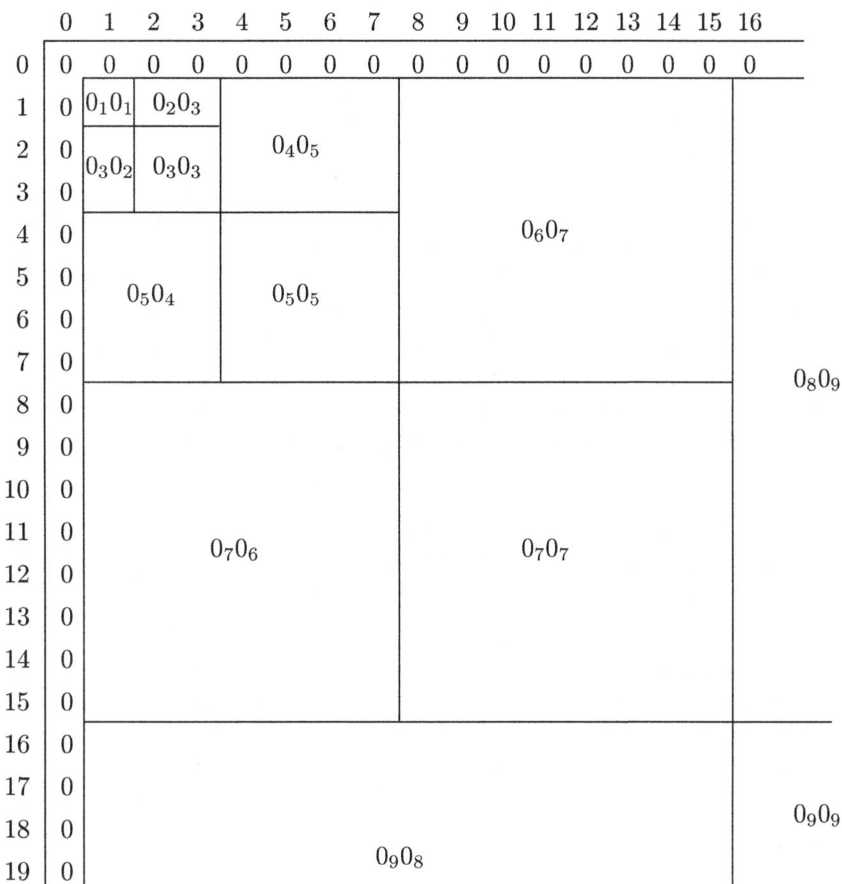

	0	1	2	3	4	5	6	7	8	9	10	11	12	13	14	15	16	
0	0	0	0	0	0	0	0	0	0	0	0	0	0	0	0	0	0	
1	0	$0_1 0_1$	$0_2 0_3$															
2	0	$0_3 0_2$	$0_3 0_3$			$0_4 0_5$												
3	0																	
4	0											$0_6 0_7$						
5	0		$0_5 0_4$			$0_5 0_5$												
6	0																	
7	0																	$0_8 0_9$
8	0																	
9	0																	
10	0																	
11	0				$0_7 0_6$							$0_7 0_7$						
12	0																	
13	0																	
14	0																	
15	0																	
16	0																	
17	0																	
18	0																	$0_9 0_9$
19	0								$0_9 0_8$									

Table 2. Tallies for Coolcakes.

Why does have the tallies $0_2 0_3$? We have

$$0_1 0_1 \vee 0_1 0_1$$

$$0_2 0_3 \quad = \quad \{0_2 \mathbf{0_3}, \qquad 0_1 \mathbf{0_1} \qquad | \quad \mathbf{0_0} 0_0\}.$$

From the right tallies 0_3 and 0_1 Left will choose his best option 0_1. Right sees only the left tally 0_0, so we try the tentative tallies

$$¿ \quad \{..0_1 \,|\, 0_0 ..\} = 0_2 0_1 \quad ?$$

However, to avoid breaking the Lukewarmth Commandment these must be corrected:

$$¡\quad \{..0_1 \mid 0_0..\} = 0_2 0_3 \quad !$$

Select Boys and Girls

When Left and Right last held a children's party they took so long working out those complicated values that all hell broke loose among their hungry guests. Now the boys and girls are carefully selected and several may be seated in a single move. Left may seat any number of new boys, or Right any number of new girls, provided that, at each move, any two new arrivals are separated by a previously seated child. Once again no boy and girl may sit in adjacent seats.

The values are intriguing (Table 3). We can work them out from the equations (very like those in Chapter 5 for Seating Boys and Girls, the disjunctive form)

$$\left.\begin{array}{rl} \mathrm{L}n\mathrm{L} = & \{\mathrm{L}a\mathrm{L} \vee \mathrm{L}b\mathrm{L} \mid \mathrm{L}a\mathrm{R} \vee \mathrm{R}b\mathrm{L}\} \\ (-\mathrm{L}n\mathrm{L} =)\ \mathrm{R}n\mathrm{R} = & \{\mathrm{R}a\mathrm{L} \vee \mathrm{L}b\mathrm{R} \mid \mathrm{R}a\mathrm{R} \vee \mathrm{R}b\mathrm{R}\} \\ (\mathrm{R}n\mathrm{L} =)\ \mathrm{L}n\mathrm{R} = & \{\mathrm{L}a\mathrm{L} \vee \mathrm{L}b\mathrm{R} \mid \mathrm{L}a\mathrm{R} \vee \mathrm{R}b\mathrm{R}\} \end{array}\right\} \quad a+b=n-1,$$

$$\mathrm{L0R} \text{ and } \mathrm{R0L} \text{ illegal.}$$

The tolls show period 5 after some exceptions in the first two rows. The timers increase at three different speeds, as shown by the separating bars: some every period (a), some every other period (b), and the rest after steadily increasing numbers of periods (c). So, despite their latest precautions, Left and Right are sure to have hot battles of erratic lengths.

Mrs. Grundy

When you played Grundy's Game in Chapter 4 (the move was to split a heap into any two smaller heaps of different sizes), you may have been irritated by all those unsplittable heaps of size 1 and 2 that are left at the end. **Mrs. Grundy** has a tidy mind and allows Left to remove isolated heaps of 1, and Right isolated heaps of 2, as additional moves, and to speed the game up a little, the players may make moves in as many heaps as they like. There is a variation in which the players may split even heaps of size 4 or more into equal parts. The successive tallies are:

For $n =$	1	2	3	4	5	6	7	8	9	10	11	12
Mrs. Grundy	1	-1	$0_1 0_1$	1	$2_1{-}1_2$	$0_3 0_1$	1	$2_1{-}1_4$	$0_5 0_1$	1	$2_1{-}1_6$	$0_7 0_1$
	╳		↓	╳		↓	╳		↓	╳		↓
variation	1	-1	$0_1 0_1$	$1_2{-}2_1$	-1	$0_1 0_3$	$1_4{-}2_1$	-1	$0_1 0_5$	$1_6{-}2_1$	-1	$0_1 0_7$

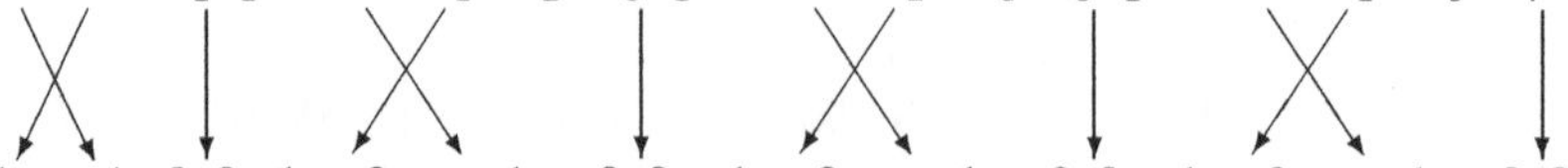

The tolls have period 3 in each case and the timers other than 1 increase steadily. The arrows display a remarkable coincidence: tallies for the variation are obtained by shuffling the negatives of those for the original.

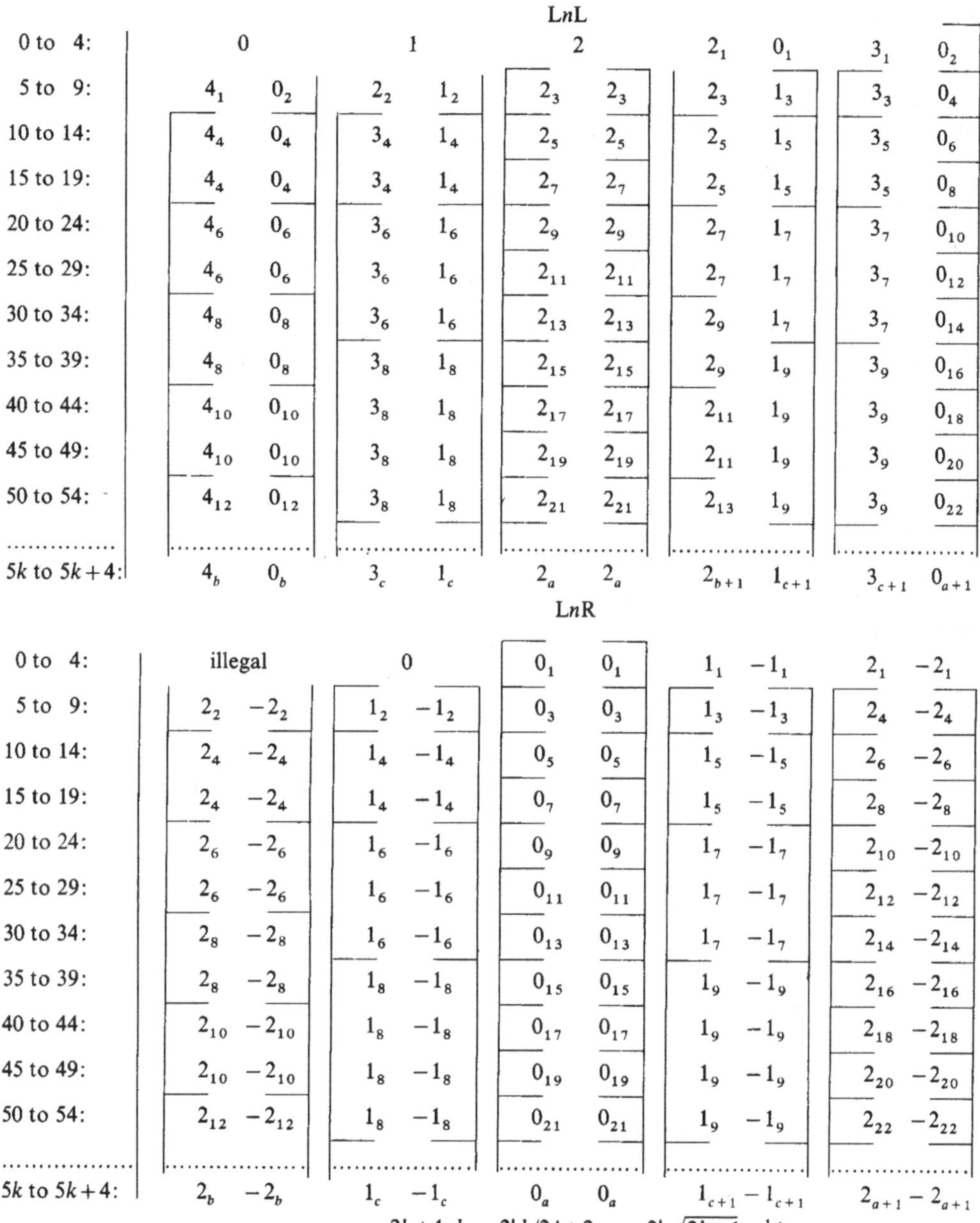

LnL

0 to 4:	0		1		2		2_1	0_1	3_1	0_2
5 to 9:	4_1	0_2	2_2	1_2	2_3	2_3	2_3	1_3	3_3	0_4
10 to 14:	4_4	0_4	3_4	1_4	2_5	2_5	2_5	1_5	3_5	0_6
15 to 19:	4_4	0_4	3_4	1_4	2_7	2_7	2_5	1_5	3_5	0_8
20 to 24:	4_6	0_6	3_6	1_6	2_9	2_9	2_7	1_7	3_7	0_{10}
25 to 29:	4_6	0_6	3_6	1_6	2_{11}	2_{11}	2_7	1_7	3_7	0_{12}
30 to 34:	4_8	0_8	3_6	1_6	2_{13}	2_{13}	2_9	1_7	3_7	0_{14}
35 to 39:	4_8	0_8	3_8	1_8	2_{15}	2_{15}	2_9	1_9	3_9	0_{16}
40 to 44:	4_{10}	0_{10}	3_8	1_8	2_{17}	2_{17}	2_{11}	1_9	3_9	0_{18}
45 to 49:	4_{10}	0_{10}	3_8	1_8	2_{19}	2_{19}	2_{11}	1_9	3_9	0_{20}
50 to 54:	4_{12}	0_{12}	3_8	1_8	2_{21}	2_{21}	2_{13}	1_9	3_9	0_{22}
....	....	....	....	....	....	....	....	....	....	....
$5k$ to $5k+4$:	4_b	0_b	3_c	1_c	2_a	2_a	2_{b+1}	1_{c+1}	3_{c+1}	0_{a+1}

LnR

0 to 4:	illegal		0		0_1	0_1	1_1	-1_1	2_1	-2_1
5 to 9:	2_2	-2_2	1_2	-1_2	0_3	0_3	1_3	-1_3	2_4	-2_4
10 to 14:	2_4	-2_4	1_4	-1_4	0_5	0_5	1_5	-1_5	2_6	-2_6
15 to 19:	2_4	-2_4	1_4	-1_4	0_7	0_7	1_5	-1_5	2_8	-2_8
20 to 24:	2_6	-2_6	1_6	-1_6	0_9	0_9	1_7	-1_7	2_{10}	-2_{10}
25 to 29:	2_6	-2_6	1_6	-1_6	0_{11}	0_{11}	1_7	-1_7	2_{12}	-2_{12}
30 to 34:	2_8	-2_8	1_6	-1_6	0_{13}	0_{13}	1_7	-1_7	2_{14}	-2_{14}
35 to 39:	2_8	-2_8	1_8	-1_8	0_{15}	0_{15}	1_9	-1_9	2_{16}	-2_{16}
40 to 44:	2_{10}	-2_{10}	1_8	-1_8	0_{17}	0_{17}	1_9	-1_9	2_{18}	-2_{18}
45 to 49:	2_{10}	-2_{10}	1_8	-1_8	0_{19}	0_{19}	1_9	-1_9	2_{20}	-2_{20}
50 to 54:	2_{12}	-2_{12}	1_8	-1_8	0_{21}	0_{21}	1_9	-1_9	2_{22}	-2_{22}
....	....	....	....	....	....	....	....	....	....	....
$5k$ to $5k+4$:	2_b	-2_b	1_c	-1_c	0_a	0_a	1_{c+1}	-1_{c+1}	2_{a+1}	-2_{a+1}

$$a = 2k+1,\ b = 2\lfloor k/2 \rfloor + 2,\ c = 2\lfloor \sqrt{2k-1} + \tfrac{1}{2} \rfloor.$$

Table 3. Tallies for Select Boys and Girls.

How to Play Misère Unions of Partizan Games

We have no idea how to play ordinary unions of partizan games with misère play.

Urgent Unions (Shotgun Weddings?)

A union of games may be played with the additional rule that it finishes as soon as its first component does. The winner of that component is then the winner of the whole game, in either normal or misère play. A component may be called finished *either* if the mover can't move *or* if neither player can move (but if we use the latter rule, we must still declare that a player loses who cannot move in any component of an unfinished game).

This kind of compound of a number of games

$$G, \quad H, \quad K, \quad \ldots$$

is called their **urgent union**

$$G \bigtriangledown H \bigtriangledown K \bigtriangledown \ldots \qquad (\text{“}G \text{ ur } H \text{ ur } K \text{ ur } \ldots\text{”}).$$

Dwight Duffus called this the *severed selective compound* (shortened selective in ONAG, and by Smith); when we wish to contrast it with the ordinary union, we call the latter the **tardy union**. The theory of urgent unions, for either normal or misère play, and with either finishing rule, follows from the more general theory given below.

Predeciders – Overriders and Suiciders

In this theory, certain moves in the components are special in that a player who makes one of them thereby instantly terminates the whole game. These **predeciding moves** are of two kinds, *overriding* and *suiciding*. A player who makes an **overriding move** wins immediately, while one who makes a **suiciding move** loses just as quickly. However, if a player cannot move when it is his turn to do so, and no predeciding move has been made, he must be said to lose.

This kind of game includes the urgent unions defined above, by making all finishing moves overriding in normal play and suiciding in misère play. But of course it also covers cases in which some finishing moves are overriding, some suiciding and some neither.

Falada

“Some of the King's Horses, First Off Wins” is a very playable game. Since it is a severed selective horse game we simply called it **Falada**. In Grimms' fairy tale, Falada was the horse whose severed head addressed the Goose Girl as she pondered her moves for the day. We play it on a quarter-infinite field with horses which move as in Chapter 9, except that at his turn a player may move however many he likes (at least one, possibly all). Horses are allowed to jump off the board and the first to do so is the winner. The game is plainly a union of component one-horse games, and it is a normal play *urgent* union because the winner of the soonest finishing component wins all.

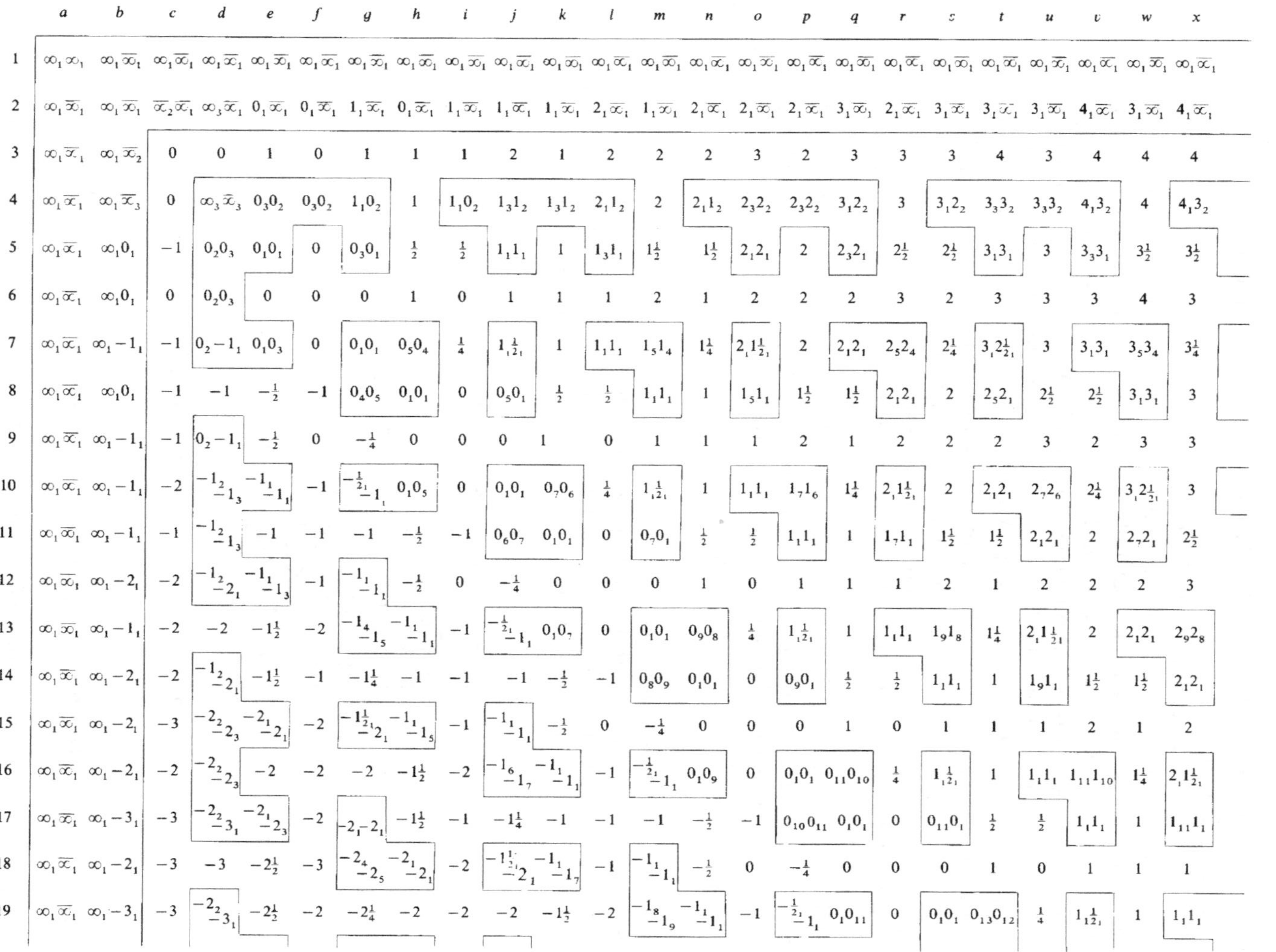

	a	b	c	d	e	f	g	h	i	j	k	l	m	n	o	p	q	r	s	t	u	v	w	x
1	$\infty_1\infty_1$	$\infty_1\overline{\infty}_1$	$\infty_1\overline{\infty}_1$	$\infty_1\overline{\infty}_1$	$\infty_1\overline{\infty}_1$	$\infty_1\overline{\infty}_1$	$\infty_1\overline{\infty}_1$	$\infty_1\overline{\infty}_1$	$\infty_1\overline{\infty}_1$	$\infty_1\overline{\infty}_1$	$\infty_1\overline{\infty}_1$	$\infty_1\overline{\infty}_1$	$\infty_1\overline{\infty}_1$	$\infty_1\overline{\infty}_1$	$\infty_1\overline{\infty}_1$	$\infty_1\overline{\infty}_1$	$\infty_1\overline{\infty}_1$	$\infty_1\overline{\infty}_1$	$\infty_1\overline{\infty}_1$	$\infty_1\overline{\infty}_1$	$\infty_1\overline{\infty}_1$	$\infty_1\overline{\infty}_1$	$\infty_1\overline{\infty}_1$	$\infty_1\overline{\infty}_1$
2	$\infty_1\overline{\infty}_1$	$\infty_1\overline{\infty}_1$	$\overline{\infty}_2\overline{\infty}_1$	$\infty_3\overline{\infty}_1$	$0_1\overline{\infty}_1$	$0_1\overline{\infty}_1$	$1_1\overline{\infty}_1$	$0_1\overline{\infty}_1$	$1_1\overline{\infty}_1$	$1_1\overline{\infty}_1$	$1_1\overline{\infty}_1$	$2_1\overline{\infty}_1$	$1_1\overline{\infty}_1$	$2_1\overline{\infty}_1$	$2_1\overline{\infty}_1$	$2_1\overline{\infty}_1$	$3_1\overline{\infty}_1$	$2_1\overline{\infty}_1$	$3_1\overline{\infty}_1$	$3_1\overline{\infty}_1$	$3_1\overline{\infty}_1$	$4_1\overline{\infty}_1$	$3_1\overline{\infty}_1$	$4_1\overline{\infty}_1$
3	$\infty_1\overline{\infty}_1$	$\infty_1\overline{\infty}_2$	0	0	1	0	1	1	1	2	1	2	2	2	3	2	3	3	3	4	3	4	4	4
4	$\infty_1\overline{\infty}_1$	$\infty_1\overline{\infty}_3$	0	$\infty_3\overline{\infty}_3$	0_30_2	0_30_2	1_10_2	1	1_10_2	1_31_2	1_31_2	2_11_2	2	2_11_2	2_32_2	2_32_2	3_12_2	3	3_12_2	3_33_2	3_33_2	4_13_2	4	4_13_2
5	$\infty_1\overline{\infty}_1$	∞_10_1	-1	0_20_3	0_10_1	0	0_30_1	$\frac{1}{2}$	$\frac{1}{2}$	1_11_1	1	1_31_1	$1\frac{1}{2}$	$1\frac{1}{2}$	2_12_1	2	2_32_1	$2\frac{1}{2}$	$2\frac{1}{2}$	3_13_1	3	3_33_1	$3\frac{1}{2}$	$3\frac{1}{2}$
6	$\infty_1\overline{\infty}_1$	∞_10_1	0	0_20_3	0	0	0	1	0	1	1	1	2	1	2	2	2	3	2	3	3	3	4	3
7	$\infty_1\overline{\infty}_1$	∞_1-1_1	-1	0_2-1_1	0_10_3	0	0_10_1	0_50_4	$\frac{1}{4}$	$1_1{\frac{1}{2}}_1$	1	1_11_1	1_51_4	$1\frac{1}{4}$	$2_11{\frac{1}{2}}_1$	2	2_12_1	2_52_4	$2\frac{1}{4}$	$3_12{\frac{1}{2}}_1$	3	3_13_1	3_53_4	$3\frac{1}{4}$
8	$\infty_1\overline{\infty}_1$	∞_10_1	-1	-1	$-\frac{1}{2}$	-1	0_40_5	0_10_1	0	0_50_1	$\frac{1}{2}$	$\frac{1}{2}$	1_11_1	1	1_51_1	$1\frac{1}{2}$	$1\frac{1}{2}$	2_12_1	2	2_52_1	$2\frac{1}{2}$	$2\frac{1}{2}$	3_13_1	3
9	$\infty_1\overline{\infty}_1$	∞_1-1_1	-1	0_2-1_1	$-\frac{1}{2}$	0	$-\frac{1}{4}$	0	0	0	1	0	1	1	1	2	1	2	2	2	3	2	3	3
10	$\infty_1\overline{\infty}_1$	∞_1-1_1	-2	$-1_2{-1_3}$	$-1_1{-1_1}$	-1	$-{\frac{1}{2}}_1{-1_1}$	0_10_5	0	0_10_1	0_70_6	$\frac{1}{4}$	$1_1{\frac{1}{2}}_1$	1	1_11_1	1_71_6	$1\frac{1}{4}$	$2_11{\frac{1}{2}}_1$	2	2_12_1	2_72_6	$2\frac{1}{4}$	$3_12{\frac{1}{2}}_1$	3
11	$\infty_1\overline{\infty}_1$	∞_1-1_1	-1	$-1_2{-1_3}$	-1	-1	-1	$-\frac{1}{2}$	-1	0_60_7	0_10_1	0	0_70_1	$\frac{1}{2}$	$\frac{1}{2}$	1_11_1	1	1_71_1	$1\frac{1}{2}$	$1\frac{1}{2}$	2_12_1	2	2_72_1	$2\frac{1}{2}$
12	$\infty_1\overline{\infty}_1$	∞_1-2_1	-2	$-1_2{-2_1}$	$-1_1{-1_3}$	-1	$-1_1{-1_1}$	$-\frac{1}{2}$	0	$-\frac{1}{4}$	0	0	0	1	0	1	1	1	2	1	2	2	2	3
13	$\infty_1\overline{\infty}_1$	∞_1-1_1	-2	-2	$-1\frac{1}{2}$	-2	$-1_4{-1_5}$	$-1_1{-1_1}$	-1	$-{\frac{1}{2}}_1{-1_1}$	0_10_7	0	0_10_1	0_90_8	$\frac{1}{4}$	$1_1{\frac{1}{2}}_1$	1	1_11_1	1_91_8	$1\frac{1}{4}$	$2_11{\frac{1}{2}}_1$	2	2_12_1	2_92_8
14	$\infty_1\overline{\infty}_1$	∞_1-2_1	-2	$-1_2{-2_1}$	$-1\frac{1}{2}$	-1	$-1\frac{1}{4}$	-1	-1	-1	$-\frac{1}{2}$	-1	0_80_9	0_10_1	0	0_90_1	$\frac{1}{2}$	$\frac{1}{2}$	1_11_1	1	1_91_1	$1\frac{1}{2}$	$1\frac{1}{2}$	2_12_1
15	$\infty_1\overline{\infty}_1$	∞_1-2_1	-3	$-2_2{-2_3}$	$-2_1{-2_1}$	-2	$-1{\frac{1}{2}}_1{-2_1}$	$-1_1{-1_5}$	-1	$-1_1{-1_1}$	$-\frac{1}{2}$	0	$-\frac{1}{4}$	0	0	0	1	0	1	1	1	2	1	2
16	$\infty_1\overline{\infty}_1$	∞_1-2_1	-2	$-2_2{-2_3}$	-2	-2	-2	$-1\frac{1}{2}$	-2	$-1_6{-1_7}$	$-1_1{-1_1}$	-1	$-{\frac{1}{2}}_1{-1_1}$	0_10_9	0	0_10_1	$0_{11}0_{10}$	$\frac{1}{4}$	$1_1{\frac{1}{2}}_1$	1	1_11_1	$1_{11}1_{10}$	$1\frac{1}{4}$	$2_11{\frac{1}{2}}_1$
17	$\infty_1\overline{\infty}_1$	∞_1-3_1	-3	$-2_2{-3_1}$	$-2_1{-2_3}$	-2	$-2_1{-2_1}$	$-1\frac{1}{2}$	-1	$-1\frac{1}{4}$	-1	-1	-1	$-\frac{1}{2}$	-1	$0_{10}0_{11}$	0_10_1	0	$0_{11}0_1$	$\frac{1}{2}$	$\frac{1}{2}$	1_11_1	1	$1_{11}1_1$
18	$\infty_1\overline{\infty}_1$	∞_1-2_1	-3	-3	$-2\frac{1}{2}$	-3	$-2_4{-2_5}$	$-2_1{-2_1}$	-2	$-1{\frac{1}{2}}_1{-2_1}$	$-1_1{-1_7}$	-1	$-1_1{-1_1}$	$-\frac{1}{2}$	0	$-\frac{1}{4}$	0	0	0	1	0	1	1	1
19	$\infty_1\overline{\infty}_1$	∞_1-3_1	-3	$-2_2{-3_1}$	$-2\frac{1}{2}$	-2	$-2\frac{1}{4}$	-2	-2	-2	$-1\frac{1}{2}$	-2	$-1_8{-1_9}$	$-1_1{-1_1}$	-1	$-{\frac{1}{2}}_1{-1_1}$	0_10_{11}	0	0_10_1	$0_{13}0_{12}$	$\frac{1}{4}$	$1_1{\frac{1}{2}}_1$	1	1_11_1

Table 4. A Filled-Out Falada Field.

Such games are solved by tallies in which the tolls are not restricted to be ordinary numbers, but are sometimes infinite.

Table 4 shows these **unrestricted tallies** for this game. A position in which one of Left's options is already an overriding win has left tally ∞_1. Another position in which *every* right option has a left tally ∞_1 will have a right tally ∞_2 and so on. In general a tally ∞_n refers to a position in which Left has a strategy which leads to an overriding win at the nth move. This will be a *left* tally if n is *odd*, a *right* one if n is *even*. A tally $-\infty_n$ (often written $\overline{\infty}_n$) is assigned to a position for which Right has an n-move overriding win strategy. It will be a *left* tally if n is even, a *right* one if n is odd. A tally ∞_0 would indicate a position that has already been (overridingly) won by Left, $\overline{\infty}_0$ one already overridingly won by Right.

Places at the very edge of the board have tallies $\infty_1\overline{\infty}_1$ since either player has an immediate overriding win; the same is true of square b2. The tallies for b3 are $\infty_1\infty_2$ since Left will win in 1 move if he starts, 2 moves if Right starts, even if Right does not move the horse on this square. But a horse on b4 can be moved by Right to c2 and so yields a 3-move win for him, corresponding to right tally $\overline{\infty}_3$. The right tally of b5 is 0_1, however, because Right will move this horse to c3, from which position neither player will voluntarily move it, lest his opponent win outright. In symbols the value of c3 is found from the equation

$$\{..\overline{\infty}_1 \mid \infty_1..\} = 0,$$

since 0 is the simplest number between $-\infty$ and $+\infty$.

An especially interesting square is d4, from which either player has an outright win in 3 moves. The play might go along one of the lines sketched in Fig. 4, but there are other alternatives and the losing player might pass on b3 or c2 by moving some other horse.

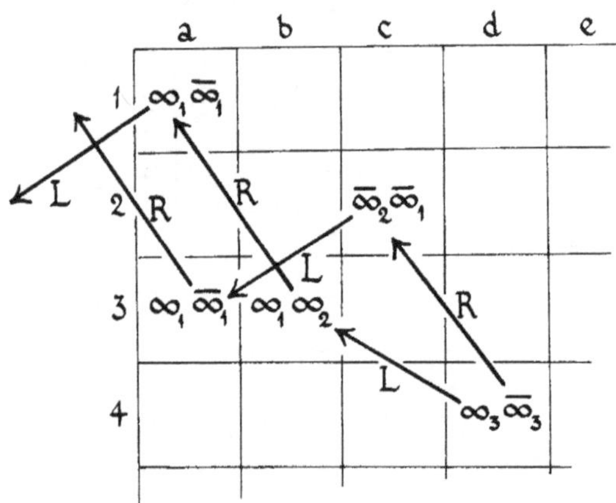

Figure 4. Two Three-Move Overriding Wins.

You'll find it hard to keep your opponents interested in Falada if you make them wait while you laboriously consult Table 4, so why not play on the beautifully patterned field of Fig. 7 in the Extras?

We'll be back with some more games after a word from the makers of Unrestricted Tallies.

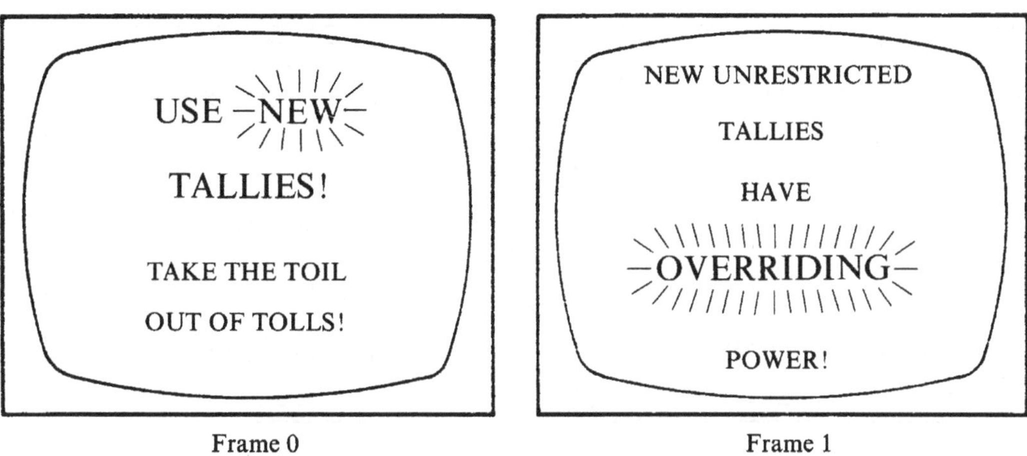

Frame 0 Frame 1

Timers on *infinite* tolls are *remoteness* functions, not suspense numbers. Therefore:

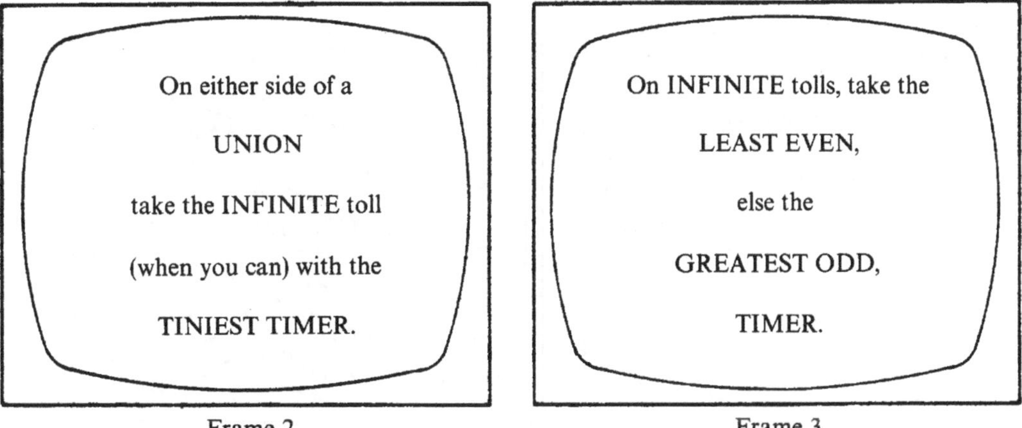

On either side of a
UNION
take the INFINITE toll
(when you can) with the
TINIEST TIMER.
On INFINITE tolls, take the
LEAST EVEN,
else the
GREATEST ODD,
TIMER.

Frame 2 Frame 3

New tallies work just as in the old machine:

FOR HOT WORK:

$$\{..\infty_a | x_b..\} = \infty_{a+1} x_{b+1},$$

$$\{..x_a | \overline{\infty}_b..\} = x_{a+1} \overline{\infty}_{b+1},$$

$$\{..\infty_a | \overline{\infty}_b..\} = \infty_{a+1} \overline{\infty}_{b+1},$$

ADD ONE to TIMERS.

Frame 4

FOR COLD WORK:

$$\{..x_a | \infty_b..\} = \{..x_a | \},$$

$$\{..\overline{\infty}_a | x_b..\} = \{ | x_b..\},$$

$$\{..\overline{\infty}_a | \infty_b..\} = 0, \text{ use the}$$

SIMPLICITY RULE.

Frame 5

But for the delicate games, whose two tallies have the same infinite toll, use the new program cycle:

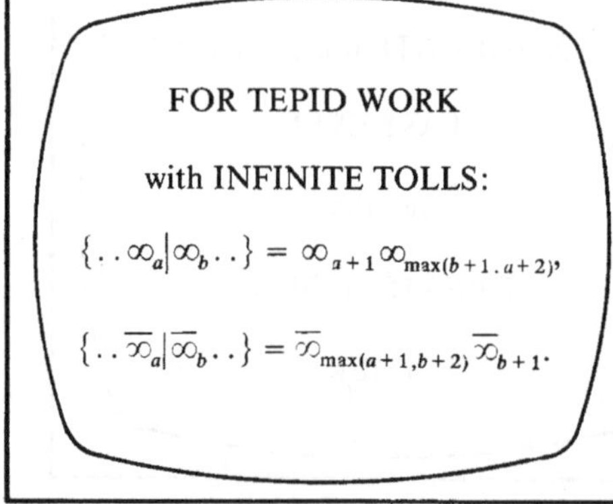

FOR TEPID WORK
with INFINITE TOLLS:
{..∞_a|∞_b..} = ∞_{a+1} ∞_{max(b+1, a+2)},
{..∞̄_a|∞̄_b..} = ∞̄_{max(a+1, b+2)} ∞̄_{b+1}.

Frame 6

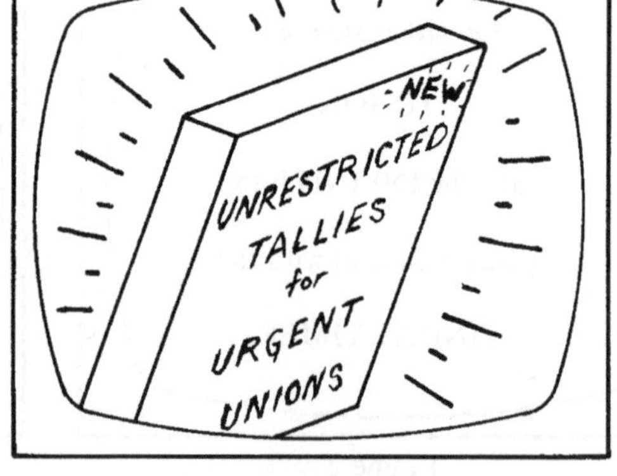

Frame 7

When using Frame 6, treat $\{..\infty_a | \}$ like $\{..\infty_a | \infty_{-1}..\}$ and $\{ | \overline{\infty}_b..\}$ like $\{..\overline{\infty}_{-1} | \overline{\infty}_b..\}$.

According to Frames 1 and 2, the tallies of

$$3_6\overline{\infty}_3 \vee \infty_9 5_{12} \vee 2\tfrac{1}{2}{}_{1}0_2 \vee \overline{\infty}_8\overline{\infty}_5$$

are

$$\overline{\infty}_8\overline{\infty}_3,$$

since the tallies with infinite tolls on the left side are ∞_9 and $\overline{\infty}_8$, and $\overline{\infty}_8$ has the tiniest timer. On the right side $\overline{\infty}_3$ beats $\overline{\infty}_5$. On *each* side the tally with infinite toll (whether ∞ or $\overline{\infty}$) and least timer overrides the others.

The timers on infinite tolls are *remoteness* functions, not suspense numbers, since a player with an *overriding* win wants to win quickly, while his opponent tries to lose slowly.

Let's use Frame 3 to find tallies for the example

$$\{..\overline{\infty}_5, ..\infty_6, ..2_2, ..\overline{\infty}_1, ..\infty_4, \mid 1_9.., 3_8.., 1_6.., \infty_1.., 1_2..\}.$$

In this Left shortlists his options with the greatest right toll (∞) and Right those with least left toll (1), to obtain

$$\{..\infty_6, ..\infty_4, \mid 1_9.., 1_6.., 1_2..\} = \{..\infty_4 \mid 1_6..\} = \infty_5 1_7.$$

Since his timers are *remoteness* functions, Left prefers ∞_4 to ∞_6 (*least* even, else *greatest* odd), but Right, who sees *suspense* numbers, prefers 1_6 to 1_9 and 1_2 (*greatest* even, else *least* odd). Since this game is hot, its tallies $\infty_5 1_7$ are found in the obvious way (Frame 4); had it been cold we should still have used the Simplicity Rule (Frame 5).

To cover the new kind of tepid case (Frame 6), when the best options for each side have the same infinite toll, we bring you a new commandment:

The *Markworthy Commandment*:

IF *INFINITE* TOLLS BE EQUAL,
THOU SHALT NOT PERMIT AN
ODD TIMER UNLESS THERE
BE A GREATER *EVEN* ONE

Thus,

$$\{..\infty_4 \mid \infty_7..\} = \infty_5\infty_8,$$

since this obeys the commandment, but

$$\{..\infty_8 \mid \infty_3..\} = \infty_9\infty_{10},$$

since the trial tallies $\infty_9\infty_4$ would disobey it. Cases when one player has no option (or upsets the board in a fit of anger) are dealt with by adjoining the suicidal option $..\overline{\infty}_{-1}$ for Left or $\infty_{-1}..$ for Right. Thus

$$\{..\infty_6 \mid \} = \{..\infty_6 \mid \infty_{-1}..\} = \infty_7\infty_8.$$

A detailed explanation for these rules will be found in the Extras, where we shall also explain why battles sometimes end before their time.

Here is what your new tally machine will look like:

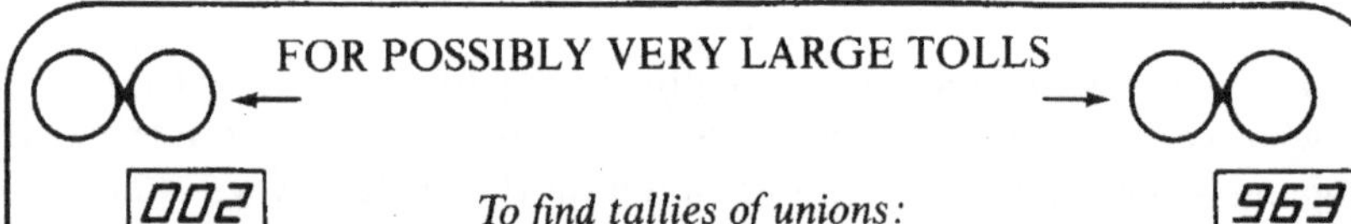

To find tallies of unions:

On each side:
If all tolls are FINITE, operate like old machine.
Otherwise, take the INFINITE TOLL with the TINIEST TIMER.

To find tallies from options:

Select best options as before, using $-\infty < x < \infty$, except that among tallies with INFINITE TOLLS, prefer that with the LEAST EVEN TIMER if there is one, otherwise that with the GREATEST ODD TIMER ("no option" counts as $\overline{\infty}_{-1}$ for Left, ∞_{-1} for Right).
From the resulting form

$$\{\,.\,.\,x_a | y_b\,.\,.\,\}$$

compute tallies as before, *except* that if $x = y$ and both are infinite, you should (if necessary) increase the *even* timer of $x_{a+1}y_{b+1}$ by just enough to make it an even number larger than the odd one.

NEW UNRESTRICTED TALLY MACHINE

One for You, Two for Me, Nothing for Both of Us

This game, also called **Squares Off**, is played with several heaps of beans. The move is to take perfect square numbers (greater than 1) of beans away from any number of heaps. Since a move removes at least 4 beans, heaps of 0, 1, 2, 3 cannot be further reduced. We declare that a move leaving 0 is an overriding win for whoever makes it (*nothing for both of us*), while one leaving 1 is an overriding win for Right (*one for you*) and one leaving 2 is an overriding win for Left (*two for me*). A move leaving 3 is not an overriding win for either player and does not terminate the game unless all other heaps are also of size 3. Table 5 gives the values.

1–10	$\overline{\infty}$	∞	0	$\infty_1\overline{\infty}_1$	$\overline{\infty}_2\overline{\infty}_1$	$\infty_1\infty_2$	0_10_1	0	$\infty_1\overline{\infty}_1$	$\infty_3\overline{\infty}_1$
11–20	∞_10_2	0_10_1	0	$\overline{\infty}_4\overline{\infty}_3$	$\infty_3\infty_4$	$\infty_1\overline{\infty}_1$	$0_1\overline{\infty}_1$	$\infty_1\overline{\infty}_5$	∞_50_1	1
21–30	$0_2\overline{\infty}_3$	∞_30_1	$0_2\overline{\infty}_5$	∞_50_1	$\infty_1\overline{\infty}_1$	$0_2\overline{\infty}_1$	∞_10_3	0_10_1	1_10_1	$\overline{\infty}_4\overline{\infty}_3$
31–40	$\infty_3\infty_4$	0_20_3	0_10_1	$\overline{\infty}_6\overline{\infty}_5$	∞_50_3	$\infty_1\overline{\infty}_1$	$0_2\overline{\infty}_1$	$\infty_1\overline{\infty}_7$	$0_1\overline{\infty}_5$	$\infty_5\infty_6$
41–50	$0_4\overline{\infty}_3$	∞_30_3	$0_2\overline{\infty}_7$	∞_70_1	1_10_5	$0_4\overline{\infty}_5$	∞_50_3	0_20_3	$\infty_1\overline{\infty}_1$	$\overline{\infty}_6\overline{\infty}_1$
51–60	∞_10_3	0_10_3	0_20_3	$0_2\overline{\infty}_3$	$\infty_3\overline{\infty}_5$	∞_50_2	0_10_5	0_20_3	$\overline{\infty}_8\overline{\infty}_7$	1
61–70	0	0_10_5	$0_4\overline{\infty}_5$	$\infty_1\overline{\infty}_1$	$\infty_7\overline{\infty}_1$	$\infty_1\overline{\infty}_5$	∞_50_5	$0_2\overline{\infty}_9$	$1_1\overline{\infty}_3$	$\infty_3\overline{\infty}_7$
71–80	0_20_5	0_30_5	0_20_3	0_40_5	$0_3\overline{\infty}_7$	∞_70_2	0_10_5	$0_2\overline{\infty}_5$	$\infty_5\overline{\infty}_5$	∞_50_3
81–90	$\infty_1\overline{\infty}_1$	$0_2\overline{\infty}_1$	$\infty_1\overline{\infty}_7$	$1_1\overline{\infty}_9$	1_10_3	$0_1\overline{\infty}_3$	∞_30_3	0_20_5	∞_70_3	0_60_7
91–100	0_40_5	0_30_3	0_20_3	$0_1\overline{\infty}_5$	$\infty_5\overline{\infty}_5$	∞_50_3	0_10_3	$0_4\overline{\infty}_7$	$0_4\overline{\infty}_7$	$\infty_1\overline{\infty}_1$

Table 5. How "Squares Off" Jumbles Finite and Infinite Tolls.

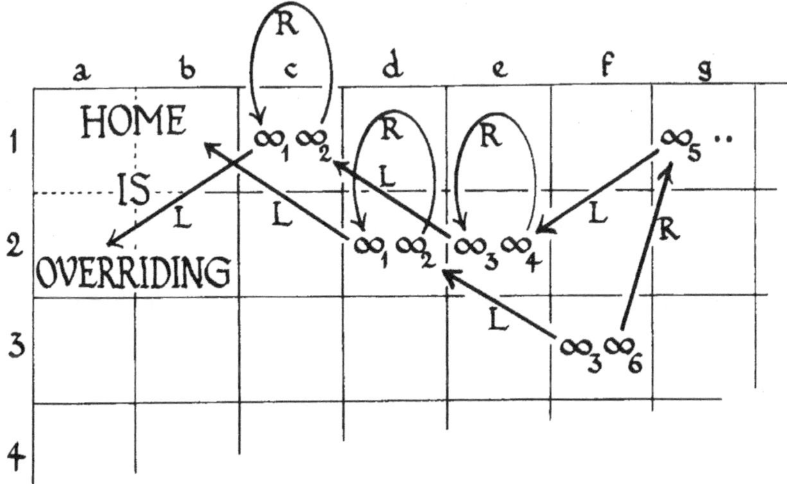

Figure 5. f3's Falada Fate.

Two More Falada Games

In each of these the players are forbidden to jump off the board and it is the first horse *home* that decides the game. Table 6 gives tallies for the first home *wins* version. Here there are two well-defined regions yielding overriding wins bordering a central region with finite tallies. Figure 5 forecasts the fate of a horse on the square f3 with tallies $\infty_3\infty_6$.

If the first horse home *loses* we get Table 7 in which no infinite tolls appear. In fact, whenever the only predeciding moves are suiciding ones, this always happens. In such cases we always get the same tallies as we would by passing a(n unenforceable?) law making suicide *illegal*.

Both the above games may end with a predeciding move. In this case the player unable, or unwilling, to move loses.

	a	b	c	d	e	f	g	h	i	j	k	l
1	Home is		$\infty_1\infty_2$	$\infty_1\infty_2$	$\infty_3\infty_4$	$\infty_3\infty_4$	$\infty_5\infty_6$	$\infty_5\infty_6$	$\infty_7\infty_8$	$\infty_7\infty_8$	$\infty_9\infty_{10}$	$\infty_9\infty_{10}$
2	Overriding!		$\infty_1\infty_2$	$\infty_1\infty_2$	$\infty_3\infty_4$	$\infty_3\infty_4$	$\infty_5\infty_6$	$\infty_5\infty_6$	$\infty_7\infty_8$	$\infty_7\infty_8$	$\infty_9\infty_{10}$	$\infty_9\infty_{10}$
3	$\bar\infty_2\bar\infty_1$	$\bar\infty_2\bar\infty_1$	$\infty_1\bar\infty_1$	$\infty_1\infty_4$	$\infty_3\infty_4$	$\infty_3\infty_6$	$\infty_5\infty_6$	$\infty_5\infty_8$	$\infty_7\infty_8$	$\infty_7\infty_{10}$	$\infty_9\infty_{10}$	$\infty_9\infty_{12}$
4	$\bar\infty_2\bar\infty_1$	$\bar\infty_2\bar\infty_1$	$\bar\infty_4\bar\infty_1$	0	0	$\infty_5\infty_6$	$\infty_5\infty_6$	$\infty_7\infty_8$	$\infty_7\infty_8$	$\infty_9\infty_{10}$	$\infty_9\infty_{10}$	$\infty_{11}\infty_{12}$
5	$\bar\infty_4\bar\infty_3$	$\bar\infty_4\bar\infty_3$	$\bar\infty_4\bar\infty_3$	0	0	1	1	$\infty_7\infty_8$	$\infty_7\infty_8$	$\infty_9\infty_{10}$	$\infty_9\infty_{10}$	$\infty_{11}\infty_{12}$
6	$\bar\infty_4\bar\infty_3$	$\bar\infty_4\bar\infty_3$	$\bar\infty_6\bar\infty_3$	$\bar\infty_6\bar\infty_5$	-1	0_10_1	1	2	2	$\infty_9\infty_{10}$	$\infty_9\infty_{12}$	$\infty_{11}\infty_{12}$
7	$\bar\infty_6\bar\infty_5$	$\bar\infty_6\bar\infty_5$	$\bar\infty_6\bar\infty_5$	$\bar\infty_6\bar\infty_5$	-1	-1	0	0	2	3	3	$\infty_{11}\infty_{12}$
8	$\bar\infty_6\bar\infty_5$	$\bar\infty_6\bar\infty_5$	$\bar\infty_8\bar\infty_5$	$\bar\infty_8\bar\infty_7$	$\bar\infty_8\bar\infty_7$	-2	0	0	1	1	3	4
9	$\bar\infty_8\bar\infty_7$	$\bar\infty_8\bar\infty_7$	$\bar\infty_8\bar\infty_7$	$\bar\infty_8\bar\infty_7$	$\bar\infty_8\bar\infty_7$	-2	-2	-1	0_10_1	1	2	2
10	$\bar\infty_8\bar\infty_7$	$\bar\infty_8\bar\infty_7$	$\bar\infty_{10}\bar\infty_7$	$\bar\infty_{10}\bar\infty_9$	$\bar\infty_{10}\bar\infty_9$	$\bar\infty_{10}\bar\infty_9$	-3	-1	-1	0	0	2
11	$\bar\infty_{10}\bar\infty_9$	$\bar\infty_{10}\bar\infty_9$	$\bar\infty_{10}\bar\infty_9$	$\bar\infty_{10}\bar\infty_9$	$\bar\infty_{10}\bar\infty_9$	$\bar\infty_{12}\bar\infty_9$	-3	-3	-2	0	0	1
12	$\bar\infty_{10}\bar\infty_9$	$\bar\infty_{10}\bar\infty_9$	$\bar\infty_{12}\bar\infty_9$	$\bar\infty_{12}\bar\infty_{11}$	$\bar\infty_{12}\bar\infty_{11}$	$\bar\infty_{12}\bar\infty_{11}$	$\bar\infty_{12}\bar\infty_{11}$	-4	-2	-2	-1	0_10_1

Table 6. Tallies for Falada, First Home Wins.

Home is		0	0	2	1	2	2	4	3	4	4	6	5	6	6
Suiciding!		1	0	1	1	3	2	3	3	5	4	5	5	7	6
0	−1	$0_1 0_1$	$0_1 0_1$	$1_1 0_1$	1	$1_1 1_1$	$1\frac{1}{2}$	$3_1 2_1$	3	$3_1 3_1$	$3\frac{1}{2}$	$5_1 4_1$	5	$5_1 5_1$	$5\frac{1}{2}$
0	0	$0_1 0_1$	0	$0_2 0_3$	$\frac{1}{2}$	$\frac{1}{2}$	2	1	2	2	4	3	4	4	6
−2	−1	$0_1 -1_1$	$0_3 0_2$	0	1	0	1	1	3	2	3	3	5	4	5
−1	−1	−1	$-\frac{1}{2}$	−1	$0_3 0_3$	$\frac{1}{4}$	$1_1 \frac{1}{2}_1$	1	$1_1 1_1$	$1\frac{1}{2}$	$3_1 2_1$	3	$3_1 3_1$	$3\frac{1}{2}$	$5_1 4_1$
−2	−3	$-1_1 -1_1$	$-\frac{1}{2}$	0	$-\frac{1}{4}$	0	$0_4 0_5$	$\frac{1}{2}$	$\frac{1}{2}$	2	1	2	2	4	3
−2	−2	$-1\frac{1}{2}$	−2	−1	$-\frac{1}{2}_1 -1_1$	$0_5 0_4$	0	1	0	1	1	3	2	3	3
−4	−3	$-2_1 -3_1$	−1	−1	−1	$-\frac{1}{2}$	−1	$0_5 0_5$	$\frac{1}{4}$	$1_1 \frac{1}{2}_1$	1	$1_1 1_1$	$1\frac{1}{2}$	$3_1 2_1$	3
−3	−3	−3	−2	−3	$-1_1 -1_1$	$-\frac{1}{2}$	0	$-\frac{1}{4}$	0	$0_6 0_7$	$\frac{1}{2}$	$\frac{1}{2}$	2	1	2
−4	−5	$-3_1 -3_1$	−2	−2	$-1\frac{1}{2}$	−2	−1	$-\frac{1}{2}_1 -1_1$	$0_7 0_6$	0	1	0	1	1	3
−4	−4	$-3\frac{1}{2}$	−4	−3	$-2_1 -3_1$	−1	−1	−1	$-\frac{1}{2}$	−1	$0_7 0_7$	$\frac{1}{4}$	$1_1 \frac{1}{2}_1$	1	$1_1 1_1$
−6	−5	$-4_1 -5_1$	−3	−3	−3	−2	−3	$-1_1 -1_1$	$-\frac{1}{2}$	0	$-\frac{1}{4}$	0	$0_8 0_9$	$\frac{1}{2}$	$\frac{1}{2}$
−5	−5	−5	−4	−5	$-3_1 -3_1$	−2	−2	$-1\frac{1}{2}$	−2	−1	$-\frac{1}{2}_1 -1_1$	$0_9 0_8$	0	1	0
−6	−7	$-5_1 -5_1$	−4	−4	$-3\frac{1}{2}$	−4	−3	$-2_1 -3_1$	−1	−1	−1	$-\frac{1}{2}$	−1	$0_9 0_9$	$\frac{1}{4}$
−6	−6	$-5\frac{1}{2}$	−6	−5	$-4_1 -5_1$	−3	−3	−3	−2	−3	$-1_1 -1_1$	$-\frac{1}{2}$	0	$-\frac{1}{4}$	0

Table 7. Tallies for Falada, First Home Loses.

Baked Alaska

This is a particularly interesting modification of Coolcakes. It is played exactly like that game, except that a person who sees a *square* cake may eat it and win the game outright. The tallies for a square cake are therefore $\infty_1\overline{\infty}_1$ but apart from this the game behaves like an ordinary union in which it is illegal to create a square cake, so that all other tolls are finite. (Many other games with predeciding moves behave like this.)

To make Table 8 as clear as possible we've written

$$a, b, c, d, e, \quad \text{instead of} \quad 1,2,3,4,5,$$

as timers, and written simply ☽ for the illegal square cakes. For a vertical strip of two squares, Left moves to 0, but Right has no (non-suicidal) move, so the value is 1. For longer vertical strips, the value is successively halved since Right's best move is to eat an end square. Other small cakes tend to be hot, but there are some surprising exceptions (61/128 for a 5 by 11 cake!) The central values are mostly cold, with a hot coating, as the name might suggest. It seems that for large enough m, the values of m by $m+1$ cakes are alternately $-1/256$ and $(2^{-8}+2^{2-m})_a 0_a$ (m odd), while those of m by $m+k$ cakes

for	$k=2$	3	4	5	6	7	8	9	10	11
are always	0	$\frac{1}{4}$	$\frac{3}{8}$	$\frac{7}{16}$	$\frac{1}{2}$	$\frac{3}{4}$	$\frac{7}{8}$	$\frac{15}{16}$	$\frac{31}{32}$	1

The 6 by 17 cake is hot with tallies

$$\frac{133}{128}{}_3\ \frac{2039}{2048}{}_5.$$

Figures 6(a) and 6(b) show which parts of this cake should remain after the hot battles started by Left and Right respectively.

Table 8. Finicky Figures for Baked Alaska.

	1	2	3	4	5	6	7	8
1	☽	-1	$-\frac{1}{2}$	$-\frac{1}{4}$	$-\frac{1}{8}$	$-\frac{1}{16}$	$-\frac{1}{32}$	
2	1	☽	$2_a-\frac{1}{2}{}_a$	$-\frac{1}{2}$	$\frac{1}{2}{}_b-\frac{1}{8}{}_a$	$\frac{1}{2}{}_a-\frac{1}{16}{}_a$	$\frac{7}{8}{}_b-\frac{1}{32}{}_a$	
3	$\frac{1}{2}$	$\frac{1}{2}{}_a-2_a$	☽	-1	$-\frac{1}{2}$	$-\frac{1}{4}$	$0_a-\frac{1}{16}{}_a$	$\frac{1}{4}{}_a-\frac{1}{32}{}_a$
4	$\frac{1}{4}$	$\frac{1}{2}$	1	☽	$\frac{5}{4}{}_a-\frac{1}{2}{}_a$	$\frac{3}{2}{}_a-\frac{1}{4}{}_a$	$2_a 0_b$	$\frac{1}{4}$
5	$\frac{1}{8}$	$\frac{1}{8}{}_a-\frac{1}{2}{}_b$	$\frac{1}{2}$	$\frac{1}{2}{}_a-\frac{5}{4}{}_a$	☽	$0_c-\frac{5}{16}{}_a$	$1_a-\frac{1}{32}{}_b$	0
6	$\frac{1}{16}$	$\frac{1}{16}{}_a-\frac{1}{2}{}_a$	$\frac{1}{4}$	$\frac{1}{4}{}_a-\frac{3}{2}{}_a$	$\frac{5}{16}{}_a 0_c$	☽	$\frac{3}{4}$	$\frac{3}{4}{}_a 0_a$
7	$\frac{1}{32}$	$\frac{1}{32}{}_a-\frac{7}{8}{}_b$	$\frac{1}{16}{}_a 0_a$	0_b-2_a	$\frac{1}{32}{}_b-1_a$	$-\frac{3}{4}$	☽	$-\frac{1}{2}$
8	$\frac{1}{64}$	$\vdots$	$\frac{1}{32}{}_a-\frac{1}{4}{}_a$	$-\frac{1}{4}$	0	$0_a-\frac{3}{4}{}_a$	$\frac{1}{2}$	☽
9	$\frac{1}{128}$	$\vdots$	$\frac{1}{64}{}_a-\frac{7}{16}{}_b$	$-\frac{7}{16}{}_c-\frac{1}{2}{}_b$	$-\frac{1}{4}$	$-\frac{1}{4}{}_a-\frac{13}{16}{}_a$	0	$0_a-\frac{33}{64}{}_a$
10	$\frac{1}{256}$		$\vdots$	$-\frac{15}{32}{}_c-\frac{3}{4}{}_b$	$-\frac{3}{8}$	$-\frac{1}{4}$	$-\frac{1}{8}$	0
11	$\frac{1}{512}$		$\vdots$	$-\frac{31}{64}{}_c-1_c$	$-\frac{61}{128}$	$-\frac{61}{128}{}_a-1_a$	$-\frac{1}{4}$	$-\frac{1}{4}{}_c-\frac{1}{4}{}_b$
12	$\vdots$			$-\frac{63}{128}{}_c-1\frac{1}{4}{}_a$		$-\frac{1}{2}$	$-\frac{3}{8}$	$-\frac{5}{16}$
13	$\vdots$			$\vdots$		$-\frac{3}{4}$	$-\frac{1}{2}$	$-\frac{1}{2}{}_a-\frac{1}{2}{}_a$
14				$\vdots$		$-\frac{7}{8}$	$-\frac{3}{4}$	$-\frac{1}{2}$
15						$-\frac{125}{128}{}_b-\frac{3}{2}{}_a$	$-\frac{7}{8}$	$-\frac{7}{8}{}_a-1_a$
16						$-\frac{63}{64}$	$-\frac{15}{16}$	$-\frac{7}{8}$
17							$-\frac{31}{32}$	$-\frac{15}{16}$
18							$-\frac{63}{64}$	$-\frac{31}{32}$
19								-1

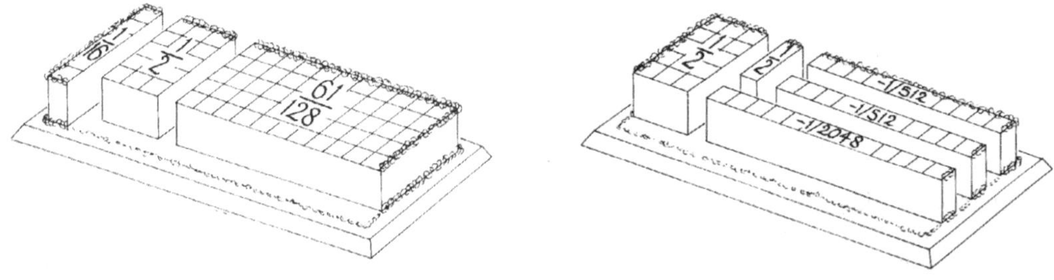

(a) First Bite to Left. (b) First Bite to Right.

Figure 6. The Frozen Remains of Baked Alaska.

9	10	11	12	13	14	15	16	17
henceforth		-2^{2-n}						
	henceforth $(1-2^{4-n})_b(-2^{2-n})_a$							
$\frac{7}{16}_b -\frac{1}{64}_a$		henceforth $(\frac{1}{2}-2^{5-n})_b(-2^{3-n})_a$						
$\frac{1}{2}_b\frac{7}{16}_c$			and, for $n \geqslant 13$, $(\frac{3}{2}-2^{9-n})_d(\frac{1}{2}-2^{5-n})_c$					
$\frac{1}{4}$	$\frac{3}{8}$	$\frac{61}{128}$	$\frac{1}{2}_a\frac{503}{1024}_b$					
$\frac{13}{16}_a\frac{1}{4}_a$	$\frac{1}{4}$	$1_a\frac{61}{128}_a$	$\frac{1}{2}$	$\frac{3}{4}$	$\frac{7}{8}$	$\frac{3}{2}_a\frac{125}{128}_b$	$\frac{63}{64}$	$\frac{133}{128}_c\frac{2039}{2048}_e$
0	$\frac{1}{8}$	$\frac{1}{4}$	$\frac{3}{8}$	$\frac{1}{2}$	$\frac{3}{4}$	$\frac{7}{8}$	$\frac{15}{16}$	$\frac{31}{32}$
$\frac{33}{64}_a 0_a$	0	$\frac{1}{4}_b\frac{1}{4}_c$	$\frac{5}{16}$	$\frac{1}{2}_a\frac{1}{2}_a$	$\frac{1}{2}$	$1_a\frac{7}{8}_a$	$\frac{7}{8}$	$\frac{15}{16}$
☽	$\frac{1}{128}_a 0_a$	0	$\frac{1}{4}$	$\frac{3}{8}$	$\frac{7}{16}$	$\frac{1}{2}$	$\frac{3}{4}$	$\frac{7}{8}$
$0_a -\frac{1}{128}_a$	☽	$-\frac{1}{256}$	0	$\frac{1}{4}$	$\frac{3}{8}$	$\frac{7}{16}$	$\frac{1}{2}$	$\frac{3}{4}$
0	$\frac{1}{256}$	☽	$\frac{3}{512}_a 0_a$	0	$\frac{1}{4}$	$\frac{3}{8}$	$\frac{7}{16}$	$\frac{1}{2}$
$-\frac{1}{4}$	0	$0_a -\frac{3}{512}_a$	☽	$-\frac{1}{256}$	0	$\frac{1}{4}$	$\frac{3}{8}$	$\frac{7}{16}$
$-\frac{3}{8}$	$-\frac{1}{4}$	0	$\frac{1}{256}$	☽	$\frac{9}{2048}_a 0_a$	0	$\frac{1}{4}$	$\frac{3}{8}$
$-\frac{7}{16}$	$-\frac{3}{8}$	$-\frac{1}{4}$	0	$0_a -\frac{9}{2048}_a$	☽	$-\frac{1}{256}$	0	$\frac{1}{4}$
$-\frac{1}{2}$	$-\frac{7}{16}$	$-\frac{3}{8}$	$-\frac{1}{4}$	0	$\frac{1}{256}$	☽	$\frac{33}{8192}_a 0_a$	0
$-\frac{3}{4}$	$-\frac{1}{2}$	$-\frac{7}{16}$	$-\frac{3}{8}$	$-\frac{1}{4}$	0	$0_a -\frac{33}{8192}_a$	☽	$-\frac{1}{256}$
$-\frac{7}{8}$	$-\frac{3}{4}$	$-\frac{1}{2}$	$-\frac{7}{16}$	$-\frac{3}{8}$	$-\frac{1}{4}$	0	$\frac{1}{256}$	☽
$-\frac{15}{16}$	$-\frac{7}{8}$	$-\frac{3}{4}$	$-\frac{1}{2}$	$-\frac{7}{16}$	$-\frac{3}{8}$	$-\frac{1}{4}$	0	$0_a -\frac{129}{32768}_a$
$-\frac{31}{32}$	$-\frac{15}{16}$	$-\frac{7}{8}$	$-\frac{3}{4}$	$-\frac{1}{2}$	$-\frac{7}{16}$	$-\frac{3}{8}$	$-\frac{1}{4}$	0

Extras

A Felicitous Falada Field

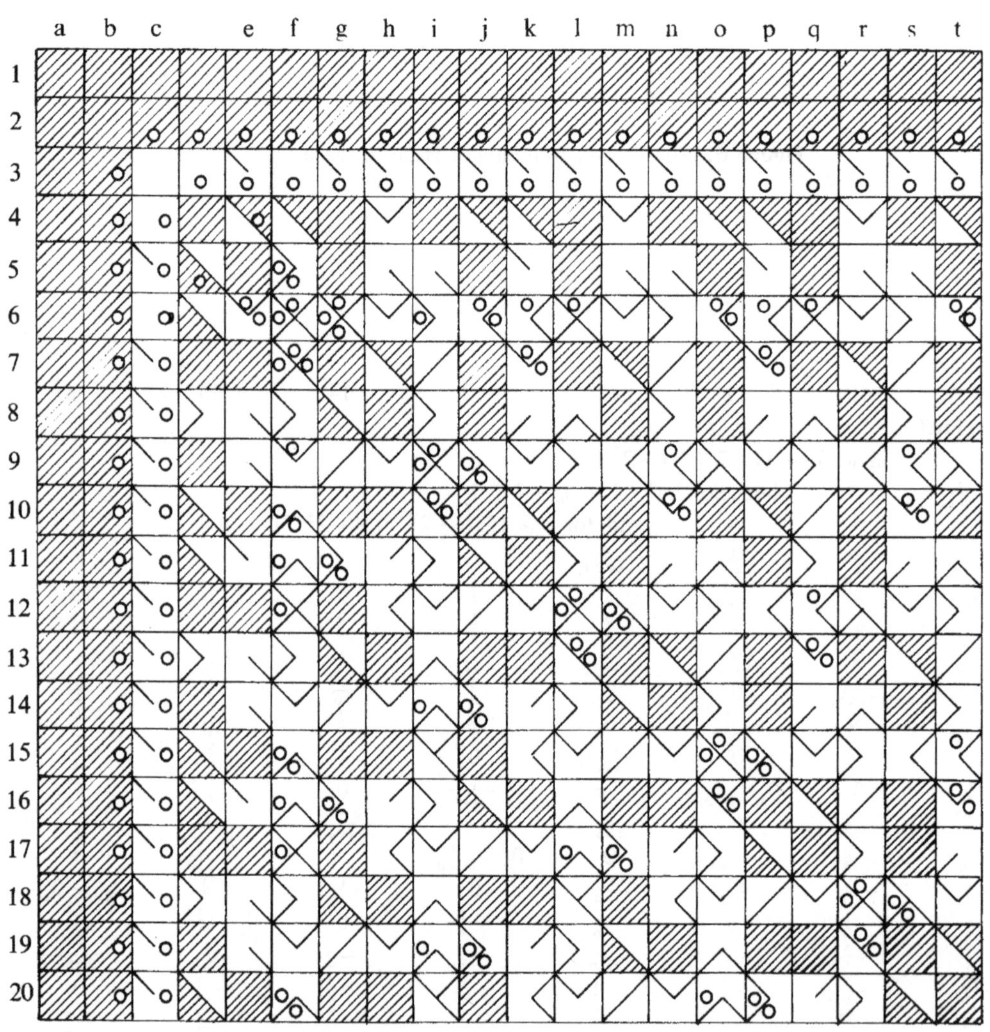

Figure 7. No Faults, Failures or Faltering on this Infallible Falada Field.

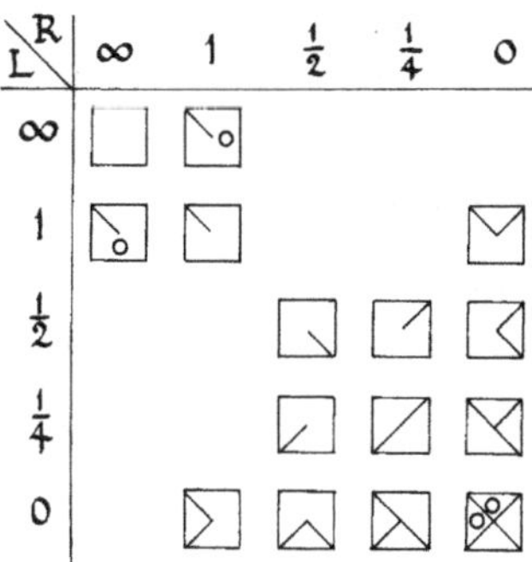

Figure 8. Toll Losses When Playing on Falada Field.

Each square should be considered as divided into Left and Right *halves* by the NW–SE diagonal. Make *normal* moves, 2 squares West and 1 square North for Left (2N & 1W for R), but *abnormal* ones, 2W and 1S for Left (2N & 1E for R) if there is a small circle in the S(E) of the square. Move every horse from any square which is shaded (in your half). If there is no such horse, move just one horse from an unshaded square containing the maximum number of half-diagonals, especially if there's a half-diagonal in your half. The key of Figure 8 goes into more detail, indicating the loss in toll value from either player's point of view. If there is no change in toll value, then the timer changes by 1, unless there is a small circle in the W(N), in which case it changes by 3, plus 2 more for each similarly placed circle found by making NW diagonal steps of length 3. For example, Right's half of square p20 is empty, so Right should hope to find a better move. The key tells us that Right's toll worsens (increases) by 1 if she moves, but Left's toll is unchanged. If he moves from the square it should be abnormally (circle in S) and his timer changes by 3 (circle in W) + 2 × 3 (squares m17, j14 and g11 contain similarly situated circles, but d8 does not) = 9. Extrapolation of Table 4 shows that Left would move from -1 to $..-1_9$, while Right would move to 0.

The Rules for Tallies on Infinite Tolls

We explain these by considering the examples

$$\{..\infty_4 \mid \infty_7..\} \quad \text{and} \quad \{..\infty_8 \mid \infty_3..\}$$

(in each of which Left is sure to win) in more detail. In the first example both players would like to move—Left to obtain his win at the 5th move and Right to postpone it to the 8th. It therefore behaves like a hot game, and has tallies $\infty_5\infty_8$. In the second example, Left should certainly move and reach his overriding win at the 9th move. But if Right chose to move in this component the game would only last 4 moves as against the 10 that would result if she made a pass move in this component and then allowed Left to take his 9-move win. She can make this pass move if she has a less harmful move elsewhere, and indeed by doing so she will win the whole game if in any component she has an overriding win in 9 moves or less.

In the case $\{..\infty_8 \mid \}$ Right will be forced to pass since she has no option and so its tallies are again $\infty_9\infty_{10}$ just as if she had $\infty_{-1}..$ for an option.

Time May Be Shorter than You Think!

Our timers sometimes give the wrong answer for the length of the hot battle, but this won't matter since they always get the winner right. For the example

$$\{..x_4 \mid x_7..\} \vee y,$$

we find the tallies

$$x_9 x_8 \vee y = (x+y)_9(x+y)_8,$$

which suggest that if Left starts, the battle will last for 9 moves. In fact it only lasts 5 moves, since to avoid moving in y he moves to

$$..x_4 \vee y$$

rather than passing in the tepid component. But since there's only one battle still being fought, and he wins it, Left doesn't care how long it lasts. The left tally x_9 indicates a battle that Left *can* drag on for 9 moves, *if he needs to.*

-11-

Games Infinite and Indefinite

"I find", said 'e, "things very much as 'ow I've always found,
For mostly they goes up and down, or else goes round and round".
Patrick Reginald Chalmers, *Green Days and Blue Days, Roundabouts and Swings.*

For ever and ever when I move.
How dull it is to pause, to make an end.
Alfred Lord Tennyson, *Ullysses*, I.21.

Infinite Games

Most of our examples in *Winning Ways* have had only finitely many positions. But a game can have infinitely many positions and still satisfy the ending condition. A few such are described in the first part of this chapter. More interesting are the games obtained by dropping the ending condition. which we call **loopy games**, since it's often possible to find oneself returning to the same position over and over again. In Chapter 12 we'll describe, amongst other things, C.A.B. Smith's complete theory for *impartial* loopy games, but in this chapter we'll see that the theory of partizan loopy games is completely different.

The ideas in this chapter are hardly used elsewhere in the book.

Infinite Hackenbush

You may have wondered why numbers like $\frac{2}{3}$ don't appear as values. The answer is, they do! To see how, let's look at some Hackenbush positions that get near $\frac{2}{3}$.

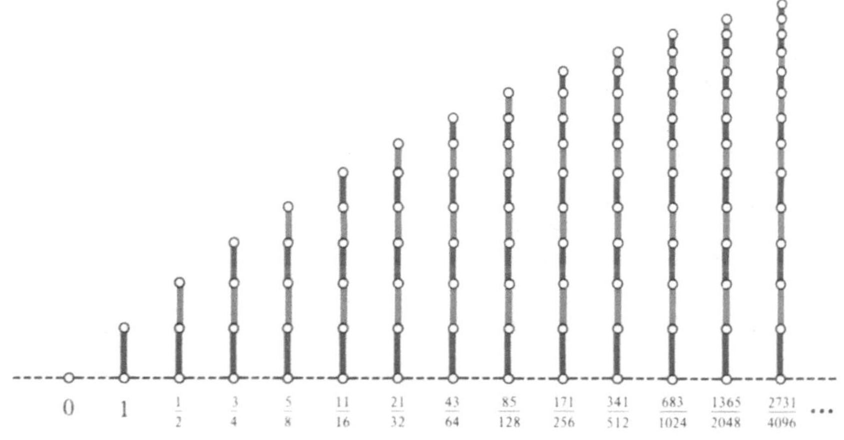

Maybe the value of an *infinite* alternating beanstalk will be *exactly* $\frac{2}{3}$? If so, Fig. 1 should be a second-player win, since its value is

$$\tfrac{2}{3} + \tfrac{2}{3} + \tfrac{2}{3} - 2 = 0.$$

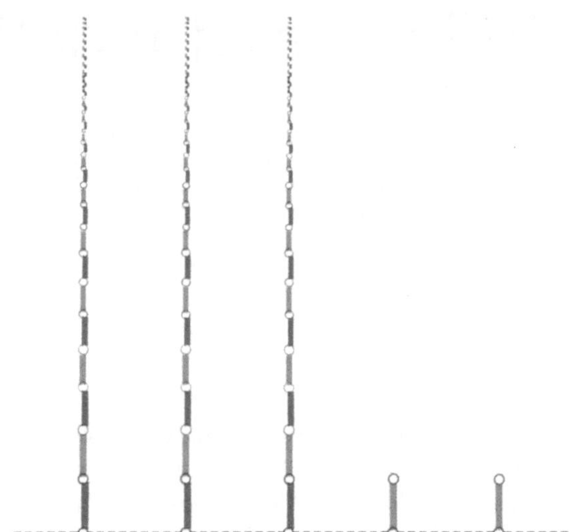

Figure 1. Two-Thirds of a Move Can Take a Long Time.

After any Left opening move the value will be

$$\left(\tfrac{2}{3} - \varepsilon\right) + \tfrac{2}{3} + \tfrac{2}{3} - 2$$

and by moving one higher up in another beanstalk, Right can achieve

$$\left(\tfrac{2}{3} - \varepsilon\right) + \left(\tfrac{2}{3} + \tfrac{1}{2}\varepsilon\right) + \tfrac{2}{3} - 2.$$

If Left now responds in the remaining beanstalk, the rest of the game is finite with value

$$\left(\tfrac{2}{3} - \varepsilon\right) + \left(\tfrac{2}{3} + \tfrac{1}{2}\varepsilon\right) + \left(\tfrac{2}{3} - \varepsilon'\right) - 2 < 0,$$

so Right wins. After any other Left move, Right can move from $\frac{2}{3}$ to $\frac{2}{3} + \frac{1}{2}\varepsilon$ leaving a finite game of value strictly less than

$$\left(\tfrac{2}{3} - \varepsilon\right) + \left(\tfrac{2}{3} + \tfrac{1}{2}\varepsilon\right) + \left(\tfrac{2}{3} + \tfrac{1}{2}\varepsilon\right) - 2 = 0.$$

It's just about as easy to check that Left wins if Right starts.

The argument is quite general and can be used to produce a Hackenbush string of any real value. For instance the binary expansion of π is

$$3.001001000011111101101010100010001000010110100 01\ldots,$$

so the string of Fig. 2, to which Berlekamp's Rule (Extras to Chapter 3) applies, has value π.

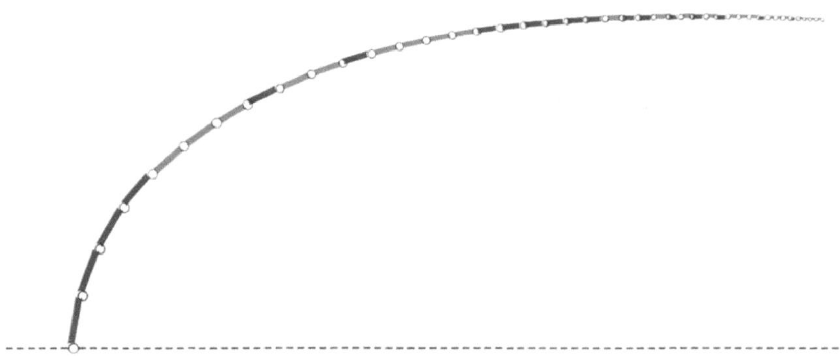

Figure 2. A Hackenbush String of Value π.

Infinite Enders

But can we really make use of Hackenbush theory for infinite graphs? Is it quite safe? Do we have to worry about limiting processes?

Yes, it's quite safe, it all works, there aren't any limiting processes involved *and* we've already proved it! Although those Hackenbush strings are infinite, any Hackenbush game you play with them will come to an end in a finite time because they satisfy the *ending condition* of Chapter 1—it's not possible to have an infinite sequence of moves (even if the players don't play alternately).

In this chapter we'll use the word **ender** for a game which satisfies this strong ending condition. You can see that it's perfectly possible for a game to have infinitely many positions and still be an ender. Although most of our examples had only finitely many positions we've been careful to make the theory apply to arbitrarily large enders.

The Infinite Ordinal Numbers

But not all infinite beanstalks have ordinary real number values. The entirely blue one marked ω in Fig. 3 has an infinite value because we can see that

$$0 < \omega, \quad 1 < \omega, \quad 2 < \omega, \quad 3 < \omega, \quad ....$$

Why didn't we just call this ∞, or maybe $\aleph_0$, since it has a countably infinite number of edges? The answer is that

$$\omega + 1 > \omega$$

whereas it's usually convenient to write

$$\infty + 1 = \infty, \ \aleph_0 + 1 = \aleph_0.$$

Figure 3 shows how, even when you've got to the top of an infinite beanstalk, you can always add another edge to make a still higher one. The kind of numbers that arise here are the **infinite ordinal numbers** studied and named by Georg Cantor and they go on and on and on and on and on and ...

$$0, 1, 2, ..., \omega, \omega+1, \omega+2, ..., \omega\times 2, ..., \omega\times 3, ..., ..., \omega\times\omega = \omega^2, ..., \omega^3, ..., \omega^4, ..., \omega^\omega, ..., \omega^{\omega^2}, ..., \omega^{\omega^\omega}, ..., ...$$

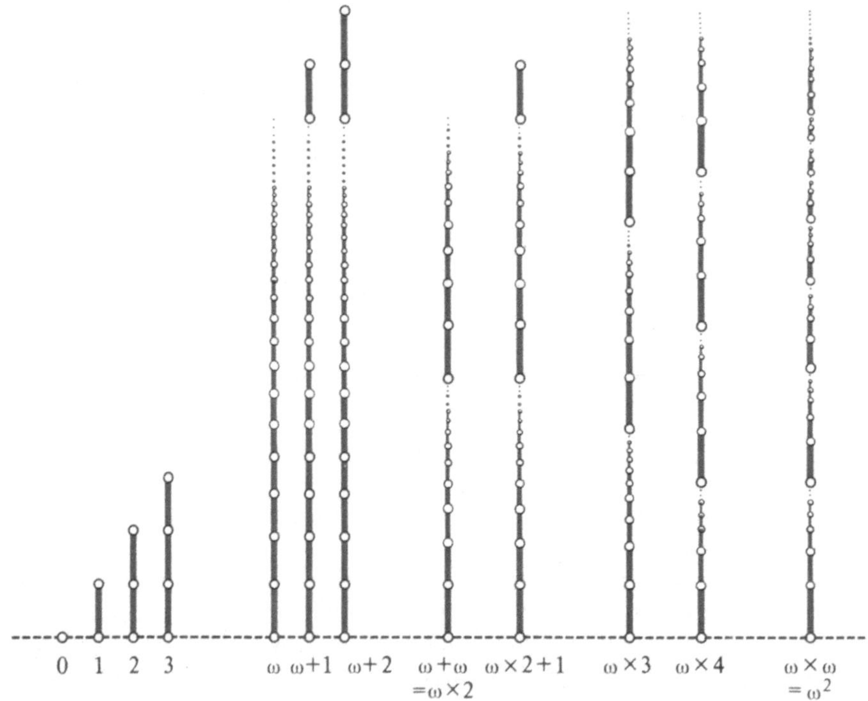

Figure 3. Beanstalks that Go On and On and On and On and On and ....

A fairly typical infinite ordinal is

$$(\omega^{\omega^\omega} \times 5) + (\omega^{\omega^2} \times 7) + \omega^\omega + (\omega^3 \times 2) + \omega^2 + 47.$$

Other Numbers

But in our theory there are even more numbers, for instance the *negatives* of Cantor's ordinal numbers, corresponding to completely *red* beanstalks, and a much greater variety when we allow blue and red edges together (Fig. 4). These are the *surreal numbers* introduced by one of us in ONAG. You can learn how to multiply them, and divide and take square roots, from ONAG; but you can *add* and *subtract* them, just by playing the games.

Beanstalks aren't the only infinite enders in Hackenbush. Figure 5 shows the values of some infinite trees. Every Blue-Red Hackenbush position that is an ender has *some* surreal number for its value (not necessarily real or finite).

Infinite Nim

You can also have purely *green* infinite beanstalks (or snakes), which are *impartial* enders. If we made all the beanstalks in Fig. 3 green, their values would include *infinite nimbers*:
$0, *1, *2, *3, ..., *\omega, *(\omega+1), *(\omega+2), ..., *(\omega\times 2), *(\omega\times 2+1), ..., *(\omega\times 3), ..., *(\omega\times 4), ..., *\omega^2, ...$

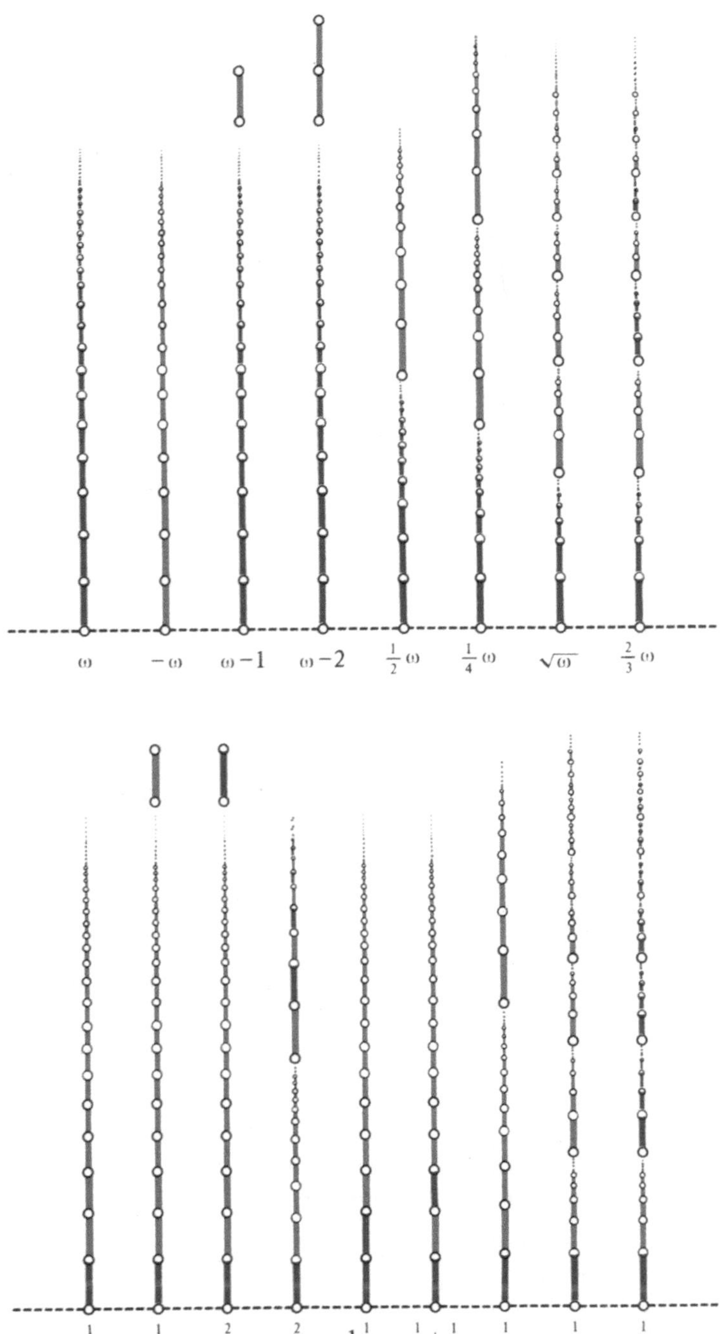

Figure 4. Brain-Baffling Bichromatic Beanstalks.

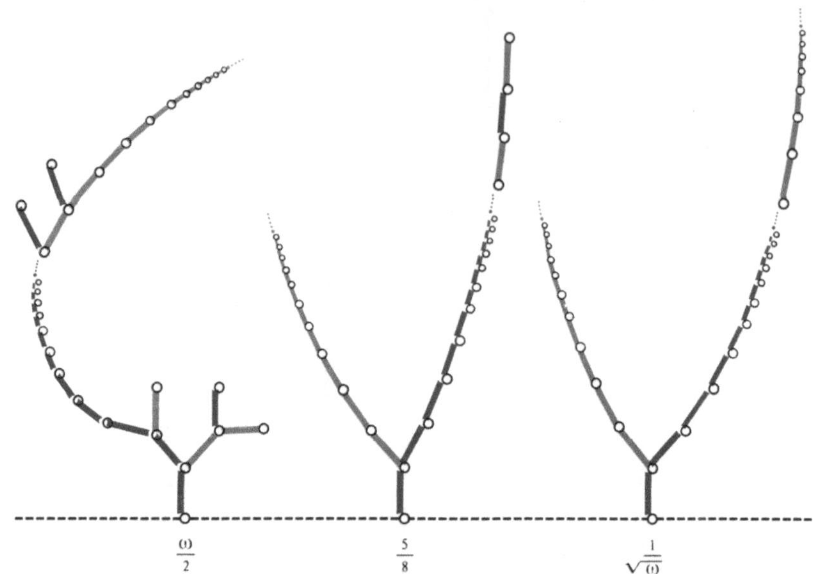

Figure 5. Some Infinite Trees (but with only *finitely* many forks).

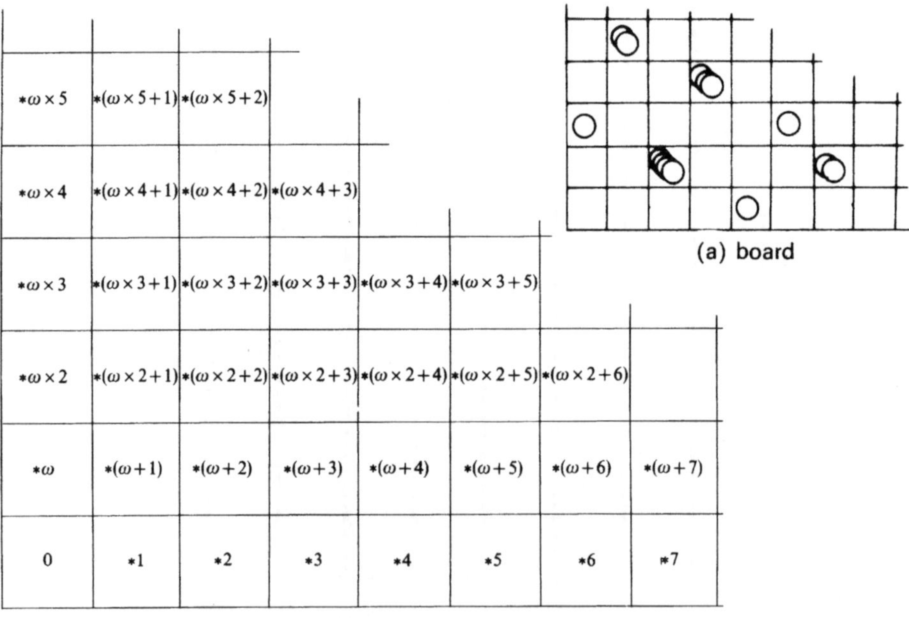

(a) board
$*\omega \times 5$ $*(\omega \times 5+1)$ $*(\omega \times 5+2)$
$*\omega \times 4$ $*(\omega \times 4+1)$ $*(\omega \times 4+2)$ $*(\omega \times 4+3)$
$*\omega \times 3$ $*(\omega \times 3+1)$ $*(\omega \times 3+2)$ $*(\omega \times 3+3)$ $*(\omega \times 3+4)$ $*(\omega \times 3+5)$
$*\omega \times 2$ $*(\omega \times 2+1)$ $*(\omega \times 2+2)$ $*(\omega \times 2+3)$ $*(\omega \times 2+4)$ $*(\omega \times 2+5)$ $*(\omega \times 2+6)$
$*\omega$ $*(\omega+1)$ $*(\omega+2)$ $*(\omega+3)$ $*(\omega+4)$ $*(\omega+5)$ $*(\omega+6)$ $*(\omega+7)$
0 $*1$ $*2$ $*3$ $*4$ $*5$ $*6$ $*7$
(b) values

Figure 6. Two-Dee Nim Has Georgous Values.

The theory of Nim applies equally to these by separately nim-adding coefficients of the various powers of ω:

$$*\left(\omega^{\omega} \times 5 + \omega^2 + \omega \times 7 + 5\right) + *\left(\omega^2 \times 3 + \omega + 3\right) = *\left(\omega^{\omega} \times 5 + \omega^2 \times 2 + \omega \times 6 + 6\right).$$

This is equivalent to expanding in powers of 2 and cancelling repetitions in pairs, when we use Cantor's equations:

$$2^{\omega} = \omega, \quad 2^{\omega \times 2} = \omega^2, \quad 2^{\omega \times 3} = \omega^3, \quad \ldots, \quad 2^{\omega^2} = \omega^{\omega}, \quad \ldots$$

(the equation $2^{\omega} = \omega$ is another reason for not using the cardinal name $\aleph_0$, which is (normally!) much smaller than $2^{\aleph_0}$).

Two-dimensional Nim uses a quarter-infinite board (Fig. 6(a)) and a finite number of counters of which any number can be on the same square. A counter may be moved any distance *leftwards* in its own row, or to *any* position in any *lower* row. The values shown in Fig. 6(b) are all the nimbers of the form

$$*(\omega \times a + b).$$

Of course, there is a three-dimensional version with typical value

$$*\left(\omega^2 \times a + \omega \times b + c\right).$$

Anyone for Hilbert Nim?

The Infinite Sprague-Grundy and Smith Theories

Any impartial ender, G, has a value of the form

$$*\alpha$$

for some, possibly infinite, ordinal number α. The value of α is found by the Mex Rule: it is the least ordinal number different from

$$\beta, \gamma, \delta, \ldots$$

where

$$*\beta, *\gamma, *\delta, \ldots$$

are the values of the options of G.

So you can use the theory of infinite Nim to play purely green Hackenbush positions, provided they're enders.

Impartial games that need not end are handled by a similar generalization of the Smith theory, which comes in Chapter 12. Their values have the forms

$$*\alpha \text{ or } \infty_{\beta\gamma\delta\ldots}$$

for possibly infinite ordinals

$$\alpha;\, \beta, \gamma, \delta, \ldots.$$

Some Superheavy Atoms

There are some quite playable games in which there arise values such as

$$\{..., -2, -1, 0, 1, 2, ...|..., -2, -1, 0, 1, 2, ...\} = \mathbb{Z}|\mathbb{Z} \text{ and } \{\mathbb{Z}|\{\mathbb{Z}|\mathbb{Z}\}\} = \mathbb{Z}||\mathbb{Z}|\mathbb{Z}.$$

(The symbol $\mathbb{Z}$ is the standard mathematical name for the collection of all integers

$$..., -3, -2, -1, 0, 1, 2, 3, \ldots.)$$

In ONAG a few of these were assigned particular names and some relationships explored, e.g.

$$\begin{aligned} \infty &= \mathbb{Z}||\mathbb{Z}|\mathbb{Z}, \\ \pm\infty &= \infty| - \infty = \mathbb{Z}|\mathbb{Z}, \\ \infty \pm \infty &= \infty|0 = \mathbb{Z}|0, \\ \infty + \infty &= 2.\infty = \mathbb{Z}||\mathbb{Z}|0 \text{ ("double infinity").} \end{aligned}$$

In many ways the games ∞ and $\pm\infty$ behave like enormously magnified versions of $\uparrow$ and $*$; there is in fact an infinitely magnifying operation defined by

$$\int^{\mathbb{Z}} G = \left\{\int^{\mathbb{Z}} G^L + n | \int^{\mathbb{Z}} G^R - n\right\}_{n=0,1,2,...}$$

and indeed we have

$$\int^{\mathbb{Z}} * = \mathbb{Z}|\mathbb{Z} = \pm\infty$$

but the integrated version of $\uparrow$ is rather smaller than $\mathbb{Z}||\mathbb{Z}|\mathbb{Z}$.

The rest of this chapter is entirely devoted to

Loopy Games

We call games that *don't* satisfy the ending condition **loopy**, because they often have closed cycles of legal moves. We will meet some impartial loopy games in Chapter 12. Now for the partizan ones!

A **play** in a loopy game G is a sequence, for example,

$$G \to G^L \to G^{LL} \to G^{LLR} \to G^{LLRL} \to G^{LLRLR} \to G^{LLRLRR} \to G^{LLRLRRR} \to ...$$

which may be finite or infinite and need not be alternately Left and Right. A play in a sum of several games defines plays in the individual components in an obvious way; for example

$$G+H \to G^L+H \to G^L+H^R \to G^{LL}+H^R \to G^{LLR}+H^R \to G^{LLR}+H^{RL} \to G^{LLRR}+H^{RL} \to G^{LLRR}+H^{RLL} \to ...$$

has the component plays

$$G \to G^L \to G^{LL} \to G^{LLR} \to G^{LLRR} \to ... \text{ in } G$$

and

$$H \to H^R \to H^{RL} \to H^{RLL} \to ... \text{ in } H.$$

Observe that when the total play is alternating, the component plays need not be.

Now when the *total* play is finite, the normal play rule tells us who wins—a player who fails to move on his turn, loses. But we must add other rules to determine the outcome when the total play is *infinite.* We shall say that

> A player *wins* the sum just if
> he wins *all* the components in
> which there is *infinite* play.
> If any infinite component play is
> *drawn*, or if two of them are won
> by *different* players, the sum is *drawn.*

Fixed, Mixed and Free

So, as well as the letter

$$\gamma, \text{ say,}$$

(we shall use loopy Greek letters for loopy games) that describes the structure (i.e. the *moves*) of any component game, we'll need something to tell us the *outcome* for each infinite play in γ. Certain infinite plays will be counted as wins for Left (+), others as wins for Right (-) and the rest as drawn games ($\pm$, or blank). A symbol

$$\gamma^\bullet$$

indicates a particular way of making these decisions, and the special cases

$$\gamma^+ \qquad \text{and} \qquad \gamma^-$$

denote the variants of γ in which *all* infinite plays are treated alike, respectively as

wins for Left and wins for Right.

If **no** infinite play is drawn, we shall say that $\gamma^\bullet$ has been **fixed**—in particular γ^+ and γ^- are the two fixed forms of γ most favorable to Left and to Right. When we omit the superscript, it will be understood that *all* infinite plays are drawn, and then we call γ **free**. Games $\gamma^\bullet$ in which some of the infinite plays are drawn, and some are not, may be called **mixed**.

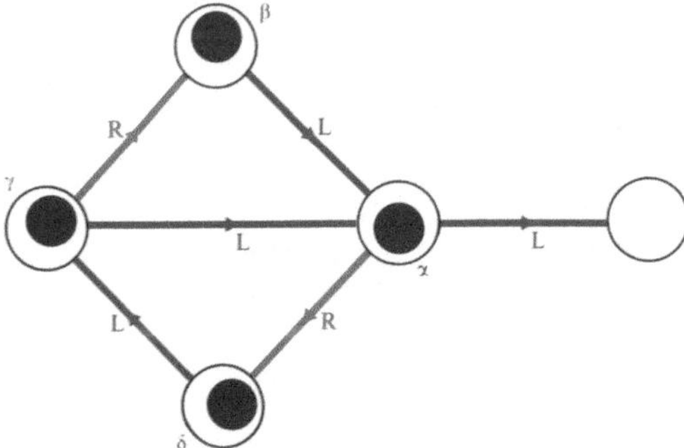

Figure 7. What's the Outcome If All Infinite Plays Are Drawn?

In Fig. 7, there's a counter on each of the positions α, β, γ and δ. Left moves by sliding a counter in the appropriate direction along a blue edge, Right along a red one. If all infinite plays are counted as draws, what's the outcome (a) if Left starts? (b) if Right starts?

Onsides and Offsides, Upsums and Downsums

We're really interested in adding free games, in which all the infinite plays are counted as drawn. Let's tell you our game plan.

In general there's a way of analyzing a loopy game γ in terms of two rather less loopy ones, s and t, and we write

$$\gamma = s\&t, \qquad s = \gamma(\mathbf{on}), \qquad t = \gamma(\mathbf{off})$$

to indicate this relationship. The games s and t are called the **sides** of γ, s being the **onside** and t the **offside**. Often the sides are ordinary ending games, or *enders*, such as $1, 4\frac{1}{2}, \uparrow, *$, and in almost all other cases they are *stoppers*, which turn out to be almost as tractable. (Always $s \geq t$.)

Now to add games given like this you'll need to know about **upsums** $(a \dotplus c)$ and **downsums** $(b \dotplus d)$, because

$$\begin{array}{lcccccc} \text{when} & \gamma & = & a & \& & b \\ \text{and} & \delta & = & c & \& & d \\ \hline \text{then} & \gamma + \delta & = & (a \dotplus c) & \& & (b \dotplus d). \end{array}$$

Use

UPSUMS for ONSIDES!
DOWNSUMS for OFFSIDES!

Onsides (and so upsums) are relevant when you intend to count draws as wins for Left, offsides (and downsums) if you count them as wins for Right. In fact loopy game theory is a sort of double-vision double-take of ending game theory (Fig. 8)!

But it's just about as easy as the ordinary theory when you know what it all means. We'll try to let you in gently ...

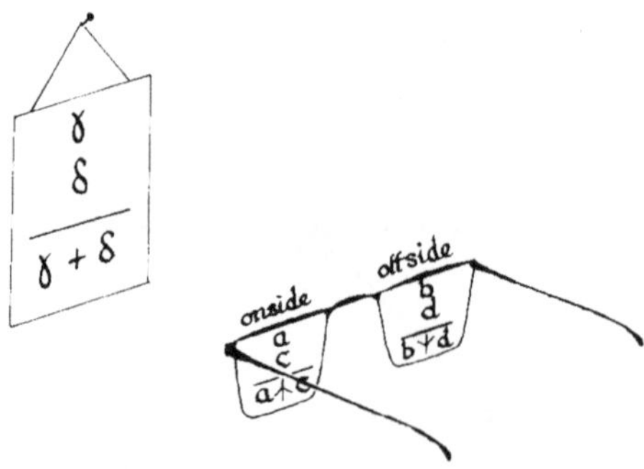

Figure 8. Seeing Both Sides of the Picture.

Stoppers

A **stopper**, g, is a game that, when played on its own, has no ultimately *alternating* (Left and Right) infinite sequence of legal moves

$$g \to g^{X} \to g^{XY} \to \ldots \to g^{XY\ldots L} \to g^{XY\ldots LR} \to g^{XY\ldots LRL} \to \ldots$$

So, *no matter which position you start playing from*, play will ultimately *stop*. (But it need not have *ended* because there might still be moves available for the player whose turn it *isn't*.)

We'll use small latin letters $(a, b, \ldots, g, \ldots, s, t, \ldots)$ for stoppers, and when we write

$$\gamma(\mathbf{on}) = s, \qquad \gamma(\mathbf{off}) = t,$$

all we mean is that s and t are stoppers for which

$$\gamma^{+} = s^{+}, \qquad \gamma^{-} = t^{-},$$

so that γ can be replaced by s if we recount all its draws as wins for Left, and by t if we recount them as wins for Right. Since then the values of s and t completely determine that of γ, we can write

$$\gamma = s\&t.$$

Notice that if γ already *is* a stopper, say ...

$$\gamma = c,$$

then

$$\gamma(\mathbf{on}) = \gamma(\mathbf{off}) = c,$$

and so

$$\gamma = c\&c.$$

But beware! Not every game has onsides and offsides which are stoppers. See Bach's Carousel in the Extras.

on, off and dud

Let's have a look at some of the simplest loopy games. You can't get much simpler than the games with only one position:

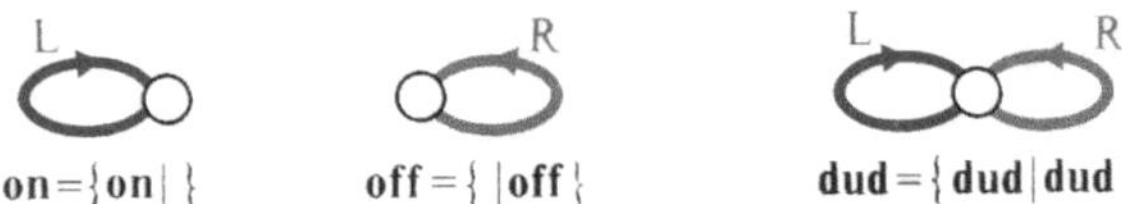

In **dud** either player can move, in **on** only Left and in **off** only Right. But in all cases the only legal moves are pass moves, back to the original position. It will turn out that

$$\mathbf{dud} = \mathbf{on} \,\&\, \mathbf{off}.$$

These values often arise in real games—some examples in Checkers appear in the Extras.

How Big Is **on**?

Well, it's not hard to see that Left wins **on** − 5, say, by moving from **on** to **on** until Right has exhausted his 5 moves. In fact, since

$$\mathbf{on} > 0, \qquad \mathbf{on} > 1, \qquad \mathbf{on} > 2, \ldots,$$

on is infinite: but by now we've met quite a few infinities—how does **on** compare with $\omega, \omega^2, \omega^\omega, \omega^{\omega^\omega}, \ldots$? The answer is:

It's Bigger than All of Them!

For if G is *any* ender, Left can win

$$\mathbf{on} - G$$

in exactly the same way. He always moves from **on** to **on**, and since G *is* an ender there must come a day when Right cannot move in it. So **on** is a sort of super-infinite number.

In mathematical logic the name **On** is used for the Class of all

Ordinal **n**umbers,

which is a kind of illegal ordinal larger than all the ordinary ones. We've adapted this name, and also decided to use **off** for its negative. You might think that

$$\mathbf{on} + \mathbf{off}$$

would be 0, but if you look at Left's and Right's options,

$$\mathbf{on} + \mathbf{off} = \{\mathbf{on} + \mathbf{off} | \mathbf{on} + \mathbf{off}\},$$

you'll see that **on** + **off** has a pass move for each player from its only position. In other words it is the game we call

dud = **d**eathless **u**niversal **d**raw,

since no matter what other game you play it with, nobody can ever bring the game to an end:

$$\mathbf{dud} + \gamma = \mathbf{dud}.$$

If any one component is **dud**, so's the sum!

Sidling Towards a Game

Here's a good way to size up (or cut down to size) the values of many loopy games. Given approximations to the values from above or below, you can put them into the equations defining your games and often get better approximations. For example, putting the inequality

$$\mathbf{on} \geq 0$$

into the definition

$$\mathbf{on} = \{\mathbf{on}|\ \}$$

we can deduce

$$\mathbf{on} \geq \{0|\ \} = 1, \qquad \text{then } \mathbf{on} \geq \{1|\ \} = 2, \ldots.$$

The Sidling Process doesn't stop with the integers! It implies

$$\mathbf{on} \geq \{0, 1, 2, ...|\ \} = \omega, \text{ then } \mathbf{on} \geq \{\omega|\ \} = \omega + 1, ..., \mathbf{on} \geq \omega^{\omega}, ...$$

once again, but without having to consider any details of strategy.

Now let's sidle in on a more complicated example. The game in Fig. 7 has five positions, $0, \alpha, \beta, \gamma, \delta$, for which the defining equations are $0 = \{\ |\ \}$,

$$\alpha = \{0|\delta\}, \qquad \beta = \{\alpha|\ \}, \qquad \gamma = \{\alpha|\beta\}, \qquad \delta = \{\gamma|\ \}.$$

How big are their values? Since we don't obviously know anything better, let us put the obvious upper bounds

$$\alpha \leq \mathbf{on}, \qquad \beta \leq \mathbf{on}, \qquad \gamma \leq \mathbf{on}, \qquad \delta \leq \mathbf{on},$$

into the defining equations. We deduce successively

$$\alpha \leq \{0|\mathbf{on}\} = 1, \qquad \beta \leq \{1|\ \} = 2, \qquad \gamma \leq \{1|2\} = 1\tfrac{1}{2}, \qquad \delta \leq \{1\tfrac{1}{2}|\ \} = 2,$$

and then

$$\alpha \leq \{0|2\} = 1.$$

This is the same as we got before, so by carrying on we'll never get any better upper bounds than

$$\alpha \leq 1, \qquad \beta \leq 2, \qquad \gamma \leq 1\tfrac{1}{2}, \qquad \delta \leq 2.$$

But we *could* start from the obvious *lower* bounds

$$\alpha \geq \mathbf{off}, \qquad \beta \geq \mathbf{off}, \qquad \gamma \geq \mathbf{off}, \qquad \delta \geq \mathbf{off},$$

and derive improved ones:

$$\beta \geq \{\mathbf{off}|\ \} = 0, \qquad \gamma \geq \{\mathbf{off}|0\} = -1, \qquad \delta \geq \{-1|\ \} = 0,$$

$$\alpha \geq \{0|0\} = *\ , \qquad \beta \geq \{*|\ \} = 0, \qquad \gamma \geq \{*|0\} = \downarrow, \qquad \delta \geq \{\downarrow|\ \} = 0.$$

The repeated value for δ shows that the process has again converged, and won't give us any better lower bounds than

$$\alpha \geq *, \qquad \beta \geq 0, \qquad \gamma \geq \downarrow, \qquad \delta \geq 0.$$

Sidling Picks Sides

What are these upper and lower bounds that the Sidling Process gives? The Sidling Theorem asserts that they are just the two sides of your game:

> Sidling in from **on** gives the *onside*,
> Sidling in from **off** gives the *offside*.

THE SIDLING THEOREM

This is proved in the Extras, but you don't have to understand the proof to be able to use the theorem. Thus in our example we found

$$\alpha \leq 1, \qquad \beta \leq 2, \qquad \gamma \leq 1\tfrac{1}{2}, \qquad \delta \leq 2,$$
$$\alpha \geq *\qquad \beta \geq 0, \qquad \gamma \geq\downarrow, \qquad \delta \geq 0,$$

and so by the Sidling Theorem we can write

$$\alpha(\mathbf{on}) = 1, \qquad \beta(\mathbf{on}) = 2, \qquad \gamma(\mathbf{on}) = 1\tfrac{1}{2}, \qquad \delta(\mathbf{on}) = 2,$$
$$\alpha(\mathbf{off}) = *\ , \qquad \beta(\mathbf{off}) = 0, \qquad \gamma(\mathbf{off}) =\downarrow, \qquad \delta(\mathbf{off}) = 0,$$

or just

$$\alpha = 1\&*, \qquad \beta = 2\&0, \qquad \gamma = 1\tfrac{1}{2}\&\downarrow, \qquad \delta = 2\&0.$$

Starting with one counter at each node in Fig. 7 gives the value

$$\alpha+\beta+\gamma+\delta=\varepsilon, \text{ say,}$$

and the Sidling Theorem tells us that ε^+ is the onside of this, namely

$$1+2+1\tfrac{1}{2}+2=6\tfrac{1}{2},$$

which is positive so that when infinite plays count as wins for Left he should win no matter who starts. But if infinite plays are wins for Right, we have ε^- which is the offside

$$*+0+\downarrow+0=\downarrow*,$$

which is fuzzy, so that whoever starts should win.

If Left starts he wins no matter what decision we make about infinite play, but if Right starts the play should continue indefinitely and the outcome is,

in ε^+,	a win for Left;
in ε^-,	a win for Right;
in ε itself,	a draw.

Stoppers Have Only One Side

We've already seen that for a stopper s the two sides are equal,

$$s = s\&s.$$

The games

over, **under**, **upon**, and **upon$*$**

are easy examples:

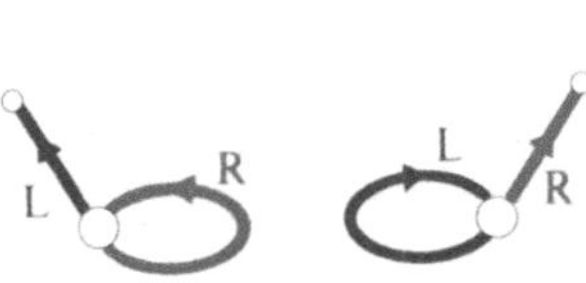

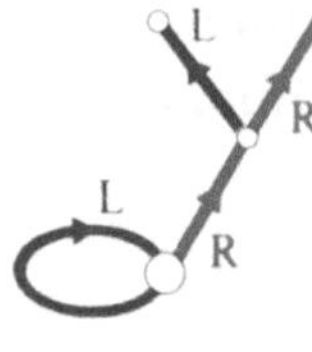

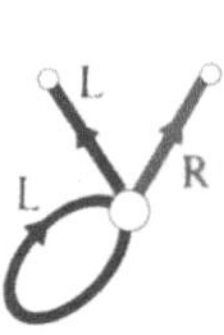

$\textbf{over} = 0|\textbf{over}$, $\textbf{under} = \textbf{under}|0$, $\textbf{upon} = \textbf{upon}|*$, $\textbf{upon}* = \{0, \textbf{upon}*|0\}$

For such games it doesn't matter which way you sidle. Thus for

$$\textbf{over} = 0|\textbf{over}$$

we get the upper bounds

$$\textbf{on},\ 0|\textbf{on} = 1,\ 0|1 = \tfrac{1}{2}, 0|\tfrac{1}{2} = \tfrac{1}{4}, ...\{0|1, \tfrac{1}{2}, \tfrac{1}{4}, ...\} = \frac{1}{\omega}, 0\left|\frac{1}{\omega} = \frac{1}{2\omega}, ...\right.$$

which suggest that

$$\textbf{over} = \frac{1}{\textbf{on}}.$$

What about sidling in from below? Since **over** is obviously positive, we can start from 0 instead of **off**, getting lower bounds

$$0, \quad 0|0 = *, \quad 0|* = \uparrow, \quad 0|\uparrow = \Uparrow *, \quad 0|\Uparrow * = \Uparrow, ...$$

which actually do tend to $\frac{1}{\textbf{on}}$ from below.

> For a *stopper*, both sidlings tend to the same place.

This saves a great deal of time when you're sidling towards stoppers because you can start wherever you like. For

$$\textbf{upon} = \textbf{upon}|*,$$

starting from 0, we get

$$0,\ 0|* = \uparrow,\ \uparrow|* = \uparrow + \uparrow^2,\ (\uparrow + \uparrow^2)|* = \uparrow + \uparrow^2 + \uparrow^3,\ ...$$

which suggest that its value is the super-infinite sum

$$\uparrow + \uparrow^2 + \uparrow^3 + ... + \uparrow^\omega + \uparrow^{\omega+1} + ... + \uparrow^{\omega\times 2} + ... + \uparrow^{\omega^2} + ...$$

(We met $\uparrow^2$ and $\uparrow^3$ in Chapter 8; $\uparrow^\alpha$ is an obvious generalization.)

Of course, since the later powers of $\uparrow$ are infinitesimal compared with $\uparrow$ itself, **upon** is still less than $\Uparrow$. We needn't do the other two examples since

$$\begin{aligned}\textbf{under} &= -\textbf{over},\\ \textbf{upon}* &= \textbf{upon} + *.\end{aligned}$$

Exercise One. Evaluate

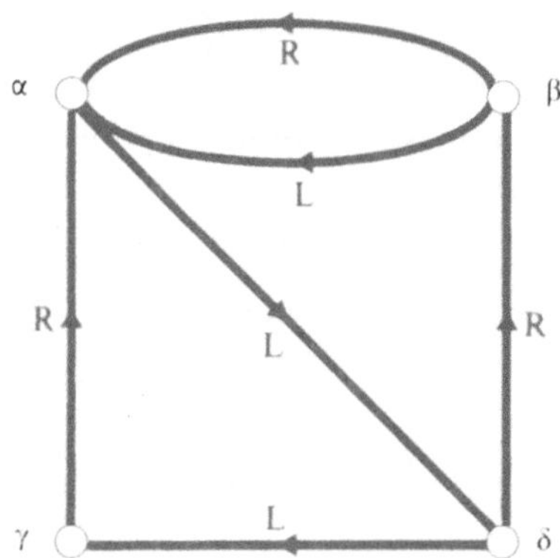

and check your answers in the Extras.

'Tis!-'Tis n!-'Tis!-'Tis n!-...

This is a well-known children's game in which Left says "Tis!", Right says "Tis n'!", Left says "Tis!" and so on, with neither player allowed two consecutive moves, as shown by the graph

or, in symbols,

$$\textbf{tis} = \{\textbf{tisn}|\}, \qquad \textbf{tisn} = \{|\textbf{tis}\}.$$

Practise your sidling and prove that

$$\textbf{tis} = 1\&0, \qquad \textbf{tisn} = 0\& - 1.$$

Given who starts, what should be the outcome of

$$\textbf{tis} + *$$

when all infinite plays are counted as draws?

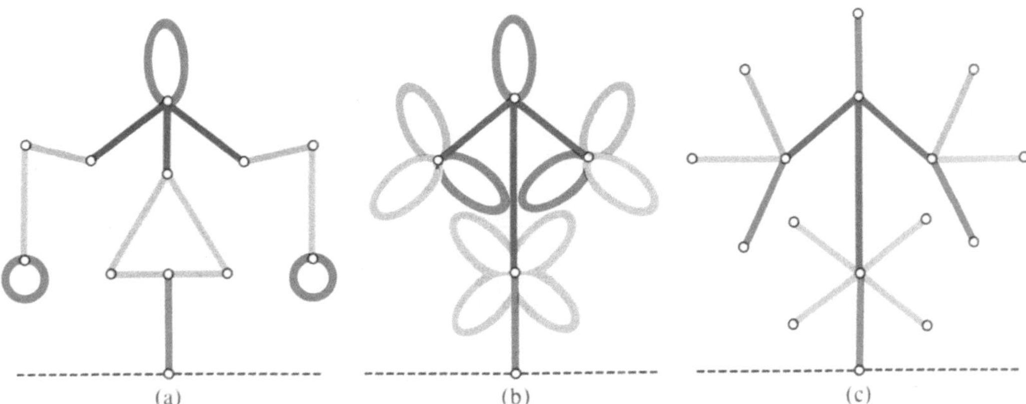

Figure 9. Girl with Yo-Yos.

Loopy Hackenbush

This game was introduced by Bob Li, who called it Double Hackenbush. As well as the normal red and blue Hackenbush edges, we have **pink** and **pale** (= pale blue) ones (Fig. 9(a)) and the new moves:

Left may repaint any *pale* edge *pink*,
Right may repaint any *pink* edge *pale*.

Pink and pale edges may not be *chopped*.

By thinking of edges that are

red, pink, absent, pale, blue

as having labels

$$-1 \qquad -\tfrac{1}{2} \qquad 0 \qquad +\tfrac{1}{2} \qquad +1$$

we can express the rules for all four types of edge by saying that

Left may *decrease* a *positive* label by 1, while
Right may *increase* a *negative* label by 1.

Disentangling Loopy Hackenbush

Since the new pink and pale edges can never be chopped, it won't affect play to bring together the two ends of any such edge, so bending it into a loop (Fig. 9(b)), and then replace such loops by *twigs* (Fig. 9(c)) just as we did when applying the Fusion Principle in Green Hackenbush (Chapter 7).

So it suffices to consider Loopy Hackenbush pictures in which all the pink and pale edges are twigs. What values have these?

Because the rules provide the following moves:

the answer is

for a *pale* twig, **tis** for a *pink* twig, **tisn**

Since

$$\mathbf{tis} = 1\&0, \qquad \mathbf{tisn} = 0\& -1,$$

this gives us a very easy way to take sides:

For *onsides*, replace *pale* by *blue* and delete *pink*. For *offsides*, replace *pink* by *red* and delete *pale*.

LI'S RULES FOR LOOPY HACKENBUSH

But beware! You must first have applied the fusion process to all pale and pink edges.

Since Li's Rules reduce Loopy Hackenbush pictures with red, pink, pale and blue edges to ordinary Blue-Red Hackenbush pictures, they show that the values have the form

$$\text{number} \ \& \ \text{number}.$$

For instance Fig. 9 has value $-\frac{1}{64}\& -\frac{5}{8}$.

Loopily Infinite Hackenbush

Infinite Hackenbush pictures can have loopy values even without pink or pale edges. For instance, let the infinite delphinium we met in Chapter 2 (Fig. 22) have value d. Then

$$d = \{0, d|0\}$$

since removing a blue petal leaves a precisely similar delphinium.

So its value is **upon***.

Several values of infinite Hackenbush pictures are shown in Fig. 10. Some of these also have pale and pink edges, but this really adds no extra generality because Li's Rules continue to apply.

Don't confuse these Hackenbush pictures with the diagrams (with arrows) showing legal moves that you, find elsewhere in this chapter.

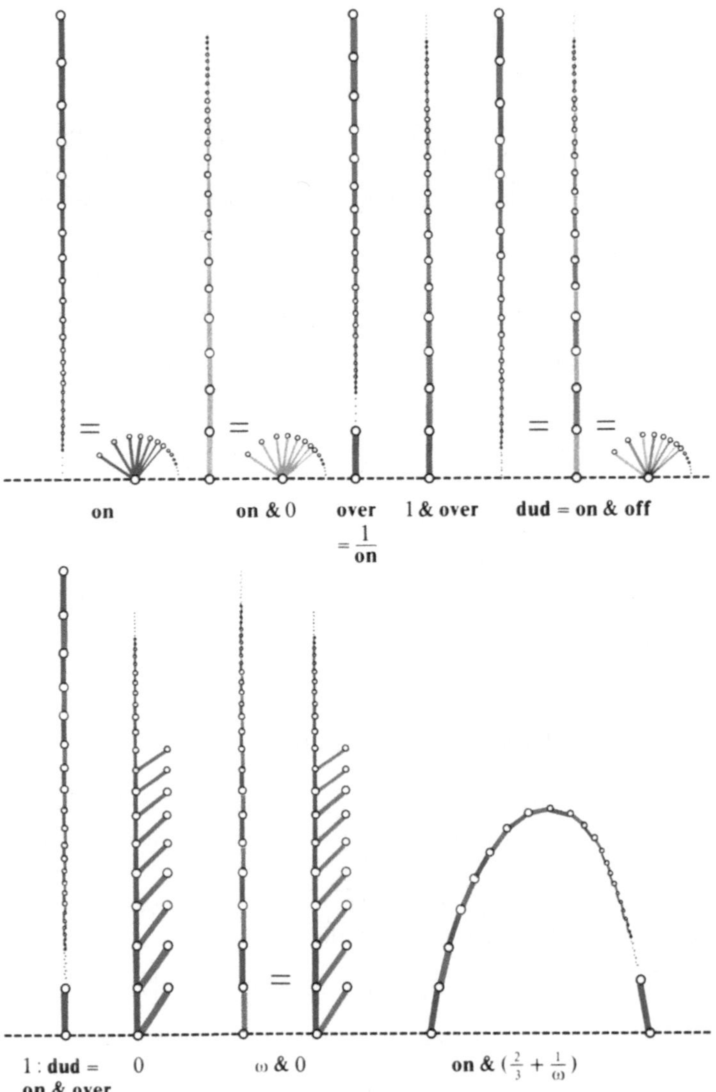

Figure 10. Large & Loopy.

Sisyphus

With useless endeavour
Forever, forever
Is Sisyphus rolling
His stone up the mountain!

Longfellow: *Masque of Pandora* (Chorus of the Eumenides).

Here's a game to practise your sidling with (Fig. 11).

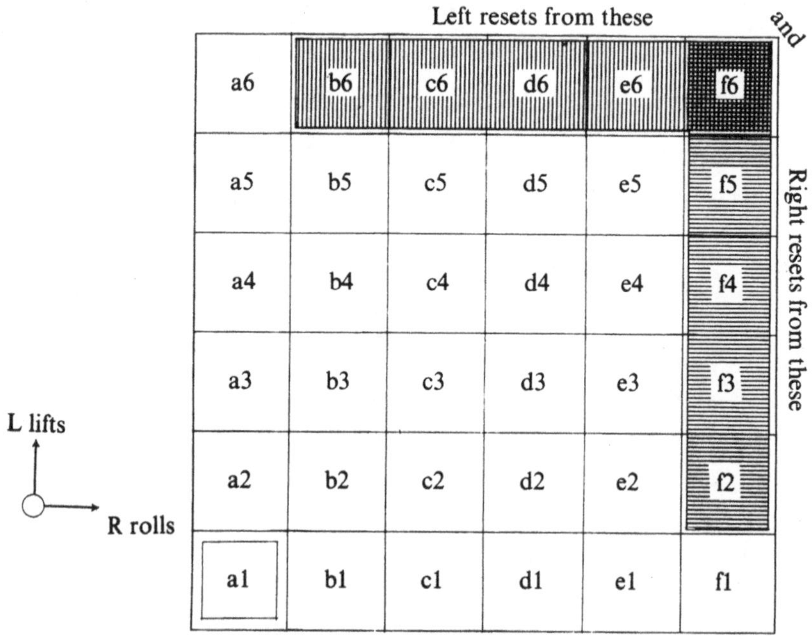

Figure 11. A Sisyphus Board.

Start with any number of stones on any of the squares. When it's his turn, Right *rolls* one stone one place to the right, Left *lifts* a stone one place upwards. Whenever there's a stone on any of the squares $\boxed{b6}$, $\boxed{c6}$, $\boxed{d6}$, $\boxed{e6}$, $\boxed{f6}$, Left has the additional option of **resetting** it to $\boxed{a1}$, and likewise a stone on one of $\boxed{f2}$, $\boxed{f3}$, $\boxed{f4}$, $\boxed{f5}$, $\boxed{f6}$ may be reset to $\boxed{a1}$ by Right.

To find the onsides we sidle down the first column:

$$\boxed{a6} \leq |\mathbf{on} = 0, \quad \boxed{a5} \leq 0|\mathbf{on} = 1, \quad \boxed{a4} \leq 1|\mathbf{on} = 2, \quad \boxed{a3} \leq 3, \quad \boxed{a2} \leq 4, \quad \boxed{a1} \leq 5,$$

and then southwest from the top right corner:

$$\boxed{f6} \leq 5|5 = 5*, \ \boxed{e6} \leq 5|5* = 5\uparrow, \ \boxed{f5} \leq 5*|5 = 5\downarrow, \ \boxed{e5} = 5\uparrow|5\downarrow = 5*$$
$$\boxed{d6} \leq 5|5\uparrow = 5\Uparrow*, \ \boxed{f4} \leq 5\downarrow|5 = 5\Downarrow*, \ldots$$

leading to the *provisional* approximations of Fig. 12.

0	5⟰*	5⤊	5⇑*	5↑	5*
1	5⤊	5⇑*	5↑	5*	5↓
2	5⇑*	5↑	5*	5↓	5⇓*
3	5↑	5*	5↓	5⇓*	5⤋
4	5*	5↓	5⇓*	5⤋	5⟱*
5	5↓	5⇓*	5⤋	5⟱*	5

Figure 12. First Attempt at Onsides of Sisyphus.

But now we get

$$\boxed{\mathrm{a1}} \le 4|5\downarrow = 4\tfrac{1}{2}, \text{ and so } \boxed{\mathrm{f6}} \le 4\tfrac{1}{2}|4\tfrac{1}{2} = 4\tfrac{1}{2}*$$

and all the 5s get replaced by $4\frac{1}{2}$s, except that the bottom row becomes

$4\frac{1}{4}$	$4\frac{1}{2}$↓	$4\frac{1}{2}$⇓*	$4\frac{1}{2}$⤋	$4\frac{1}{2}$	5

since

$$\boxed{\mathrm{f1}} \le \{4\tfrac{1}{2} ⤋ * \,|\} = 5, \;\ldots,\; \boxed{\mathrm{a1}} \le \{4|4\tfrac{1}{2}\} = 4\tfrac{1}{4}.$$

We now find ourselves successively replacing $4\frac{1}{2}$s by $4\frac{1}{4}$s, then by $4\frac{1}{8}$s and so on, until eventually the process converges to Fig. 13.

0	$4\frac{1}{32}$⟰*	$4\frac{1}{32}$⤊	$4\frac{1}{32}$⇑*	$4\frac{1}{32}$↑	$4\frac{1}{32}$*
1	$4\frac{1}{32}$⤊	$4\frac{1}{32}$⇑*	$4\frac{1}{32}$↑	$4\frac{1}{32}$*	$4\frac{1}{32}$↓
2	$4\frac{1}{32}$⇑*	$4\frac{1}{32}$↑	$4\frac{1}{32}$*	$4\frac{1}{32}$↓	$4\frac{1}{32}$⇓*
3	$4\frac{1}{32}$↑	$4\frac{1}{32}$*	$4\frac{1}{32}$↓	$4\frac{1}{32}$⇓*	$4\frac{1}{32}$⤋
4	$4\frac{1}{32}$*	$4\frac{1}{32}$↓	$4\frac{1}{32}$⇓*	$4\frac{1}{31}$⤋	$4\frac{1}{32}$⟱*
$4\frac{1}{32}$	$4\frac{1}{16}$	$4\frac{1}{8}$	$4\frac{1}{4}$	$4\frac{1}{2}$	5

Figure 13. The Onsides of Sisyphus.

Of course, the offsides are found by swapping rows with columns and negating the values, for instance

$$\boxed{\text{d1}} = 4\tfrac{1}{4}\& -2, \boxed{\text{d2}} = 4\tfrac{1}{32} \Downarrow *\&(-4\tfrac{1}{32}) \Downarrow *.$$

Living with Loops

With what you know so far you should be able to evaluate most of the loopy games you meet in practice. If you want more examples, look near the end of the chapter. Until then we'll show you some more ideas which experts can use to simplify the calculations.

Comparing Loopy Games

Because the sides of loopy games are not always enders, we need ways of comparing them. This is one problem that's best solved for *fixed* (no draw) games.

If $\alpha^\bullet$ and $\beta^\bullet$ are *fixed* games, then
$\alpha^\bullet \geq \beta^\bullet$
just if Left can arrange to win or draw in
$\alpha^\bullet - \beta^\bullet$
provided that Right starts.

THE INEQUALITY RULE

So the condition is that Left **avoids loss** or **survives** in $\alpha^\bullet - \beta^\bullet$, supposing always that Right starts.

Now any play of $\alpha^\bullet - \beta^\bullet$ yields component plays in $\alpha^\bullet$ and $-\beta^\bullet$, and so a mirror-image play in $+\beta^\bullet$. Examination of Table 1 shows that Left's strategy must arrange that

Left has a move after every move of Right's	(i)(REMAIN)
and the resulting plays in $\alpha^\bullet$ and $\beta^\bullet$ have	
$\text{sign}(\alpha^\bullet) \geq \text{sign}(\beta^\bullet)$	(ii)(ON TOP)

SURVIVAL CONDITIONS

$\text{sign}(\alpha^\bullet)$	+	+	+	0	0	0	−	−	−
$\text{sign}(-\beta^\bullet)$	+	0	−	+	0	−	+	0	−
Does Left avoid loss?	Yes	Yes	Yes	Yes	Yes, if (i)	No	Yes	No	No
$\text{sign}(+\beta^\bullet)$	−	0	+	−	0	+	−	0	+
Is $\text{sign}(\alpha^\bullet) \geq \text{sign}(\beta^\bullet)$?	Yes	Yes	Yes	Yes	Yes	No	Yes	No	No

Table 1. The Significance of Signs.

In the table, $+$ and $-$ denote infinite plays that are respectively wins for Left and Right, while 0 denotes a *finite* play. (Because the theory works only for *fixed* games, no infinite play is *drawn*.) For unfixed games $\alpha_1^{\bullet}$ and $\alpha_2^{\bullet}$,

$$\alpha_1^{\bullet} \geq \alpha_2^{\bullet}$$

means just that both

$$(\alpha_1^{\bullet})^+ \geq (\alpha_2^{\bullet})^+ \qquad \text{and} \qquad (\alpha_1^{\bullet})^- \geq (\alpha_2^{\bullet})^-$$

where $(\alpha_1^{\bullet})^+$ and $(\alpha_1^{\bullet})^-$ are the fixed games obtained from $\alpha_1^{\bullet}$ by redefining drawn plays in $\alpha_1^{\bullet}$ as wins for Left and Right respectively.

The Inequality Rule amounts to a *definition* of inequalities between loopy games and must be shown to work. For example we must show that

$$\alpha_1^{\bullet} \geq \alpha_2^{\bullet} \qquad \textit{implies} \qquad \alpha_1^{\bullet} + \beta^{\bullet} \geq \alpha_2^{\bullet} + \beta^{\bullet},$$

where we might as well suppose that $\alpha_1^{\bullet}$, $\alpha_2^{\bullet}$ and $\beta^{\bullet}$ are fixed.

In other words, we must produce a strategy for Left in

$$\alpha_1^{\bullet} + \beta^{\bullet} - \alpha_2^{\bullet} - \beta^{\bullet}$$

that makes

$$\text{sign}(\alpha_1^{\bullet} + \beta^{\bullet})^+ \geq \text{sign}(\alpha_2^{\bullet} + \beta^{\bullet})^+$$

and

$$\text{sign}(\alpha_1^{\bullet} + \beta^{\bullet})^- \geq \text{sign}(\alpha_2^{\bullet} + \beta^{\bullet})^-$$

(note that $\alpha_1^{\bullet} + \beta^{\bullet}$ and $\alpha_2^{\bullet} + \beta^{\bullet}$ need *not* be fixed).

Fortunately this happens if Left plays his given strategy in $\alpha_1^{\bullet} - \alpha_2^{\bullet}$ and the Tweedledum and Tweedledee Strategy in $\beta^{\bullet} - \beta^{\bullet}$, because each column of the Tables

		sign $(\beta^{\bullet})$		
		$-$	0	$+$
sign $(\alpha_i^{\bullet})$	$-$	$-$	$-$	$+$
	0	$-$	0	$+$
	$+$	$+$	$+$	$+$
for		sign $(\alpha_i^{\bullet} + \beta^{\bullet})^+$		

and

		sign $(\beta^{\bullet})$		
		$-$	0	$+$
sign $(\alpha_i^{\bullet})$	$-$	$-$	$-$	$-$
	0	$-$	0	$+$
	$+$	$-$	$+$	$+$
		sign $(\alpha_i^{\bullet} + \beta^{\bullet})^-$		

increases downwards.

The Swivel Chair Strategy

We must also show that

$$\alpha^{\bullet} \geq \beta^{\bullet} \text{ and } \beta^{\bullet} \geq \gamma^{\bullet} \text{ imply } \alpha^{\bullet} \geq \gamma^{\bullet}$$

where once again we can suppose $\alpha^{\bullet}$, $\beta^{\bullet}$, $\gamma^{\bullet}$ to be fixed games.

In other words. Left is given surviving strategies in $\alpha^\bullet - \beta^\bullet$ and $\beta^\bullet - \gamma^\bullet$ and desires one in $\alpha^\bullet - \gamma^\bullet$. To find it he employs two tables, a swivel chair and Right's manservant, mr. read (Fig. 14). He plays his given strategies in the pairs $\alpha^\bullet$, $-\beta^\bullet$ and $\beta^\bullet$, $-\gamma^\bullet$ and instructs mr. read to play the Tweedledum and Tweedledee Strategy, responding to a Left move in $\beta^\bullet$ or $-\beta^\bullet$ with the Right mirror-image reply (read's lower case initial marks his humble status.)

Figure 14. The Swivel Chair Strategy.

If he does this the resulting plays will have

$$\mathrm{sign}(\alpha^\bullet) \geq \mathrm{sign}(\beta^\bullet) \text{ and } \mathrm{sign}(\beta^\bullet) \geq sign(\gamma^\bullet)$$

(mr. read's instructions make $\mathrm{sign}(\beta^\bullet) = -\mathrm{sign}(-\beta^\bullet)$), so we've certainly satisfied the ON TOP condition. Also, if the total play in the real game $\alpha^\bullet - \gamma^\bullet$ is finite, then

$$0 = \mathrm{sign}(\alpha^\bullet) \geq \mathrm{sign}(\beta^\bullet) \geq \mathrm{sign}(\gamma^\bullet) = 0$$

and so the play in all four components is finite and Left made the last move. This wasn't in $\beta^\bullet$ or $-\beta^\bullet$ since mr. read would have obediently replied, and so must have been in the real game $\alpha^\bullet - \gamma^\bullet$.

Stoppers Are Nice

We've already seen that

> a stopper s has only one side:
> $s(\mathbf{on} = s(\mathbf{off}) = s$,
> $s = s \;\&\; s$

and so

> for stoppers, both sidlings tend to the same place.

Here are some other ways in which stoppers behave like ordinary enders:

> If s and t are stoppers,
> then the inequalities
> $s^+ \geq t^+$, $s^+ \geq t^-$, $s^- \geq t^-$
> are all equivalent and
> any one of them implies
> $s \geq t$

INEQUALITIES FOR STOPPERS

This means that you can usually forget about those $^+$ and $^-$ signs for stoppers. In particular:

> To check an inequality
> $s \geq t$
> between stoppers, it suffices
> to show that Left can respond
> to every move in $s - t$,
> Right starting.

COMPARING STOPPERS

> You can simplify stoppers
> by omitting dominated options
> and bypassing reversible ones
> just as you do for enders.

SIMPLIFYING STOPPERS

> A stopper which has only finitely
> many positions has a unique
> **simplest form** with no dominated
> or reversible moves from any position.

THE SIMPLEST FORM THEOREM FOR STOPPERS

We only sketch the proofs.

If s and t are stoppers, and we have infinite play in $s-t$, then we must have infinite play in *both* s and t, for otherwise there'd be an ultimately alternating infinite sequence of moves in one of them. This establishes the inequality rules.

Now let s be any stopper and $\hat{s}$ be obtained by omitting some dominated Left options from various positions of s, retaining, of course, enough options to dominate them. If t is another stopper, and $s \geq t$, we shall show that $\hat{s} \geq t$.

What we shall show is that when Right makes any move from a position

$$\hat{s}_0 - t_0 \text{ of } \hat{s} - t$$

for which we had

$$s_0 \geq t_0$$

then Left can reply to a position

$$\hat{s}_m - t_m, \text{ say,}$$

for which we still have

$$s_m \geq t_m.$$

If Left can keep on making just one more move like this he can keep on forever!

After Right's move from

$$\hat{s}_0 - t_0$$

we arrive at a position

$$\hat{s}_1 - t_1$$

with Left to move. What did he do from the corresponding position

$$s_1 - t_1$$

of $s-t$? The moves in $-t_1$ are still available to him, so we might as well suppose he moved to

$$s_2 - t_1$$

for some Left option s_2 of s_1; now s_2 might be one of the Left options we omitted in deriving $\hat{s}$ from s, but we certainly retained some Left option s_3 of s_1 dominating s_2, and Left can move to

$$\hat{s}_3 - t_1.$$

Since we had $s_2 \geq t_1$, we must have $s_3 \geq t_1$ and Left has survived for one more move.

This shows that $\hat{s} \geq s$ and it is obvious that $\hat{s} \leq s$. A similar argument shows that we can bypass reversible options. The remaining proofs are very like those for ordinary enders.

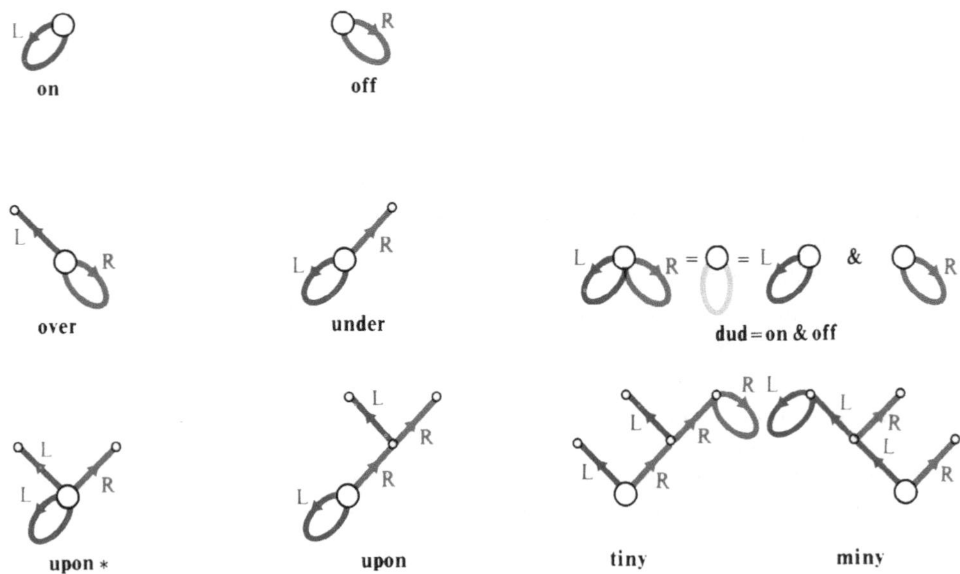

Figure 15. Some Choice Plums.

Plumtrees Are Nicer!

The moves of an ordinary ender form a tree in which blue branches correspond to Left moves and red ones to Right ones, and we may use green branches for moves which may be made by either player.

We get **plumtrees** from ordinary trees by adding plums! We use blue plums to indicate pass moves for Left, red plums for pass moves for Right and green ones for pass moves available to either player. Figure 15 shows our old examples and a new pair, **tiny** and **miny**.

How big is **tiny**?

We've met some *ordinary* tiny games

$$+_G = \{0||0| - G\} \qquad (\text{"tiny } G\text{"})$$

before. A 4×2 rectangle in Domineering has value $+_2$ and in Toads and Frogs we met $+_{\frac{1}{4}}$, and $+_{1|\frac{1}{2}}$, and found them to be very small positive games. But

$$\textbf{tiny} = +_{\textbf{on}}$$

is no ordinary game: since **on** exceeds every other game, $+_{\textbf{on}}$ is smaller than any other positive game.

tiny is the smallest positive value there is!

Of course,

$$\textbf{miny} = -_{\textbf{on}} = -\textbf{tiny}$$

is the least negative game there is.

Taking Care of Plumtrees

When all your games are plumtrees, then everything in your garden's lovely, because there are so many things you can do with them.

GATHERING. Of course you'll be keen to gather the fruits of your labors. With plumtrees this is easy because the sum of a number of plumtrees is always another plumtree, which has a pass for a given player in some position only when one of the components does.

RIPENING. We can suppose that each plumtree has at most one plum per node, since a pair of red and blue plums can be replaced by a single green one. In this form the two sides of a plumtree are easily found by ripening it:

To find the *onside*, replace all green plums by *blue* ones.
To find the *offside*, replace all green plums by *red* ones.

PRUNING. *Ripe* plumtrees (with at most one plum per node) are stoppers and so, if they have finitely many positions, can be put into simplest form by omitting dominated options and bypassing reversible ones.

GRAFTING. If g and h are ripe plumtrees with $g \geq h$, then $g\&h$ can be represented by a plumtree obtained by grafting g onto h as in Fig. 16. However, we must allow a Right move to **dud** from every g^L and a Left move to **dud** from every h^R to make sure the graft is not rejected.

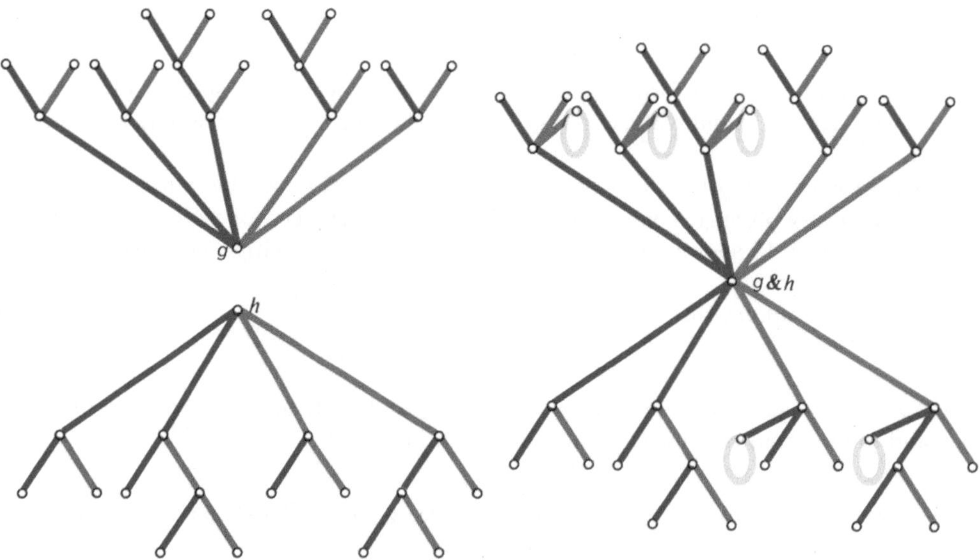

Figure 16. Grafting Two Plumtrees.

PLACING. It's often very easy to compare plumtrees with ordinary enders, using Norton's Lemma (Chapter 7, Extras), since

g:**on** is g with a blue plum added to its initial node,
g:**off** is g with a red plum added to its initial node.

For example, you can see from Fig. 15 that

$$\mathbf{upon}{*} \quad \text{is} \quad {*} : \mathbf{on},$$

and it therefore has the same order relation with g as $*$ does for every g not involving $*$. In fact

$$\mathbf{upon}{*} = {\uparrow}{*} : \mathbf{on}$$

(since $* : 1 = {\uparrow}{*}$ and $1 : \mathbf{on} = \mathbf{on}$) and so **upon**$*$ has the same order relation with g as ${\uparrow}{*}$ does, for every g not involving ${\uparrow}{*}$.

Working With Upsums and Downsums

Let's use these ideas to add two simple plumtrees.

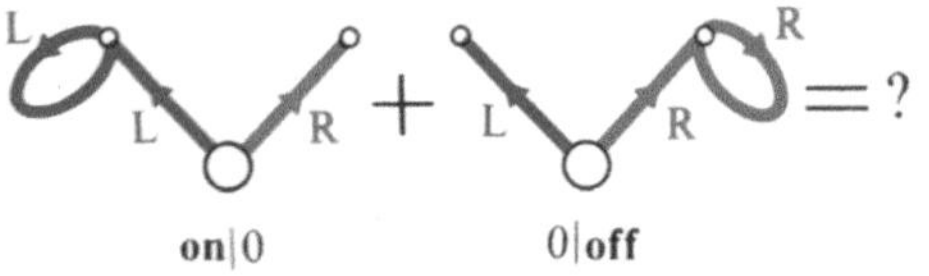

$\mathbf{on}|0$ $\qquad$ $0|\mathbf{off}$

We find

$$? = \{\ \mathbf{on} + \{0|\mathbf{off}\}, \quad \{\mathbf{on}|0\} + 0 \quad | \quad 0 + \{0|\mathbf{off}\}, \quad \{\mathbf{on}|0\} + \mathbf{off}\ \}.$$

Next

$$\mathbf{on} + \{0|\mathbf{off}\} = \{\text{pass}, \ \ \mathbf{on} + 0 | \mathbf{on} + \mathbf{off}\}$$

where the pass indicates that Left has a pass move in the sum corresponding to his pass move in **on**. The Left option **on** in this dominates the pass move and we have

$$\mathbf{on} + \mathbf{off} = \mathbf{dud} = \mathbf{on}\ \&\ \mathbf{off},$$

so that the onside of $\mathbf{on} + \{0|\mathbf{off}\}$ is

$$\mathbf{on}|\mathbf{on} = \mathbf{on},$$

while the offside is

$$\mathbf{on}|\mathbf{off} = \mathbf{hot}, \ \text{say},$$

the hottest game of all! These are the *upsums* and *downsums*:

$$\mathbf{on} \mathbin{\hat{+}} \{0|\mathbf{off}\} = \mathbf{on}, \qquad \mathbf{on} \mathbin{\check{+}} \{0|\mathbf{off}\} = \mathbf{hot},$$

or, more briefly,

$$\mathbf{on} + \{0|\mathbf{off}\} = \mathbf{on}\ \&\ \mathbf{hot},$$

and similarly

$$\{\mathbf{on}|0\} + \mathbf{off} = \mathbf{hot}\ \&\ \mathbf{off}.$$

So

$$? = \{\mathbf{on\&hot}, \{\mathbf{on}|0\} | \{0|\mathbf{off}\}, \mathbf{hot\&off}\}$$

whose onside is

$$\{\mathbf{on}, \{\mathbf{on}|0\} | \{0|\mathbf{off}\}, \mathbf{hot}\} = \{\mathbf{on} | \{0|\mathbf{off}\}\}$$

since $\{\mathbf{on}|0\}$ and $\mathbf{hot} = \{\mathbf{on}|\mathbf{off}\}$ are dominated options.

We conclude that

$$\{\mathbf{on}|0\} + \{0|\mathbf{off}\} = \{\mathbf{on}|\{0|\mathbf{off}\}\} = \mathbf{hi}, \text{ say,}$$

and similarly

$$\{\mathbf{on}|0\} + \{0|\mathbf{off}\} = \{\{\mathbf{on}|0\}|\mathbf{off}\} = \mathbf{lo}, \text{ say.}$$

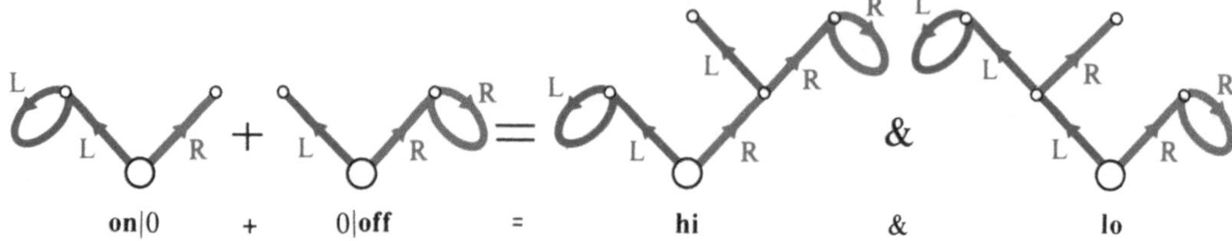

These answers are in simplest form.

on, off and hot

It's easy to add these three games to any other. We call a stopper **half-on** if it has **on** as a Left option and **half-off** if it has **off** as a Right one (the half-on games are just those of the form $\{\mathbf{on}|a, b, c, ...\}$).

Both **on** and **hot** are half-on; both **hot** and **off** are half-off; **hot** is the only half-on *and* half-off game. Here's an addition table:

	on	other half-on	**hot**	other half-off	**off**
on	**on**	**on**	**on** & **hot**	**on** & **hot**	**on** & **off**
other half-on	**on**	**on**	**on** & **hot**	?	**hot** & **off**
hot	**on** & **hot**	**on** & **hot**	**on** & **off**	**hot** & **off**	**hot** & **off**
other half-off	**on** & **hot**	?	**hot** & **off**	**off**	**off**
off	**on** & **off**	**hot** & **off**	**hot** & **off**	**off**	**off**

and here's a table for some related games (**ono** = **on**|0, **oof** = 0|**off**):

	ono	**hi**	**lo**	**oof**	**tiny** & **miny**
ono	**on**	**on** & **ono**	**ono** & **hot**	**hi** & **lo**	**on**|**tiny** & **ono**
hi	**on** & **ono**	**on** & **hi**	**hi** & **lo**	**hot** & **oof**	**hi** & **on**||**miny**|**off**
lo	**ono** & **hot**	**hi** & **lo**	**lo** & **off**	**oof** & **off**	**on**|**tiny**||**off** & **lo**
oof	**hi** & **lo**	**hot** & **oof**	**oof** & **off**	**off**	**oof** & **miny**|**off**

and here's a picture of some order relations (Fig. 17).

So

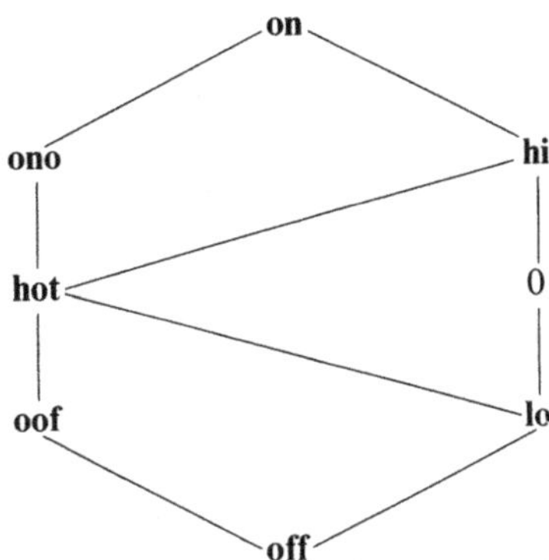

Figure 17. Higher or Lower?

A Summary of Some Sum Properties

Upsums and downsums of stoppers have properties like ordinary sums and there are a few other relations between them and the ordering properties. Our mathematical readers might like a list:

$0 \mathbin{+\!\uparrow} a = a$	$0 \mathbin{+\!\downarrow} a = a$	Zero Law
$a \mathbin{+\!\uparrow} b = b \mathbin{+\!\uparrow} a$	$a \mathbin{+\!\downarrow} b = b \mathbin{+\!\downarrow} a$	Commutativity
$(a \mathbin{+\!\uparrow} b) \mathbin{+\!\uparrow} c = a \mathbin{+\!\uparrow} (b \mathbin{+\!\uparrow} c)$	$(a \mathbin{+\!\downarrow} b) \mathbin{+\!\downarrow} c = a \mathbin{+\!\downarrow} (b \mathbin{+\!\downarrow} c)$	Associativity
$-(a \mathbin{+\!\uparrow} b) = (-a) \mathbin{+\!\downarrow} (-b)$	$-(a \mathbin{+\!\downarrow} b) = (-a) \mathbin{+\!\uparrow} (-b)$	Negation
$a' \geq a \Longrightarrow a' \mathbin{+\!\uparrow} b \geq a \mathbin{+\!\uparrow} b$	$a' \geq a \Longrightarrow a' \mathbin{+\!\downarrow} b \geq a \mathbin{+\!\downarrow} b$	Monotonicity
$a \mathbin{+\!\uparrow} b \geq a \mathbin{+\!\downarrow} b$		a, b Property
$a \mathbin{+\!\downarrow} b \geq c$ just if $a \geq (-b) \mathbin{+\!\uparrow} c$		a, b, c Property

Some of the consequences of these are quite surprising, as we'll see later.

The House of Cards

Look at this list of games:

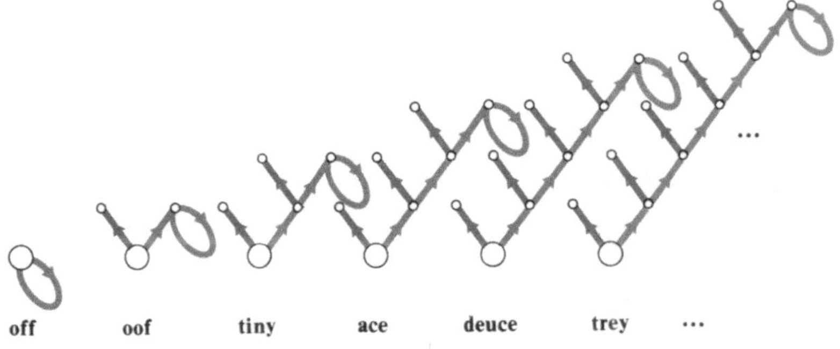

Figure 18. Plumtrees in the Uplands.

How big are they?

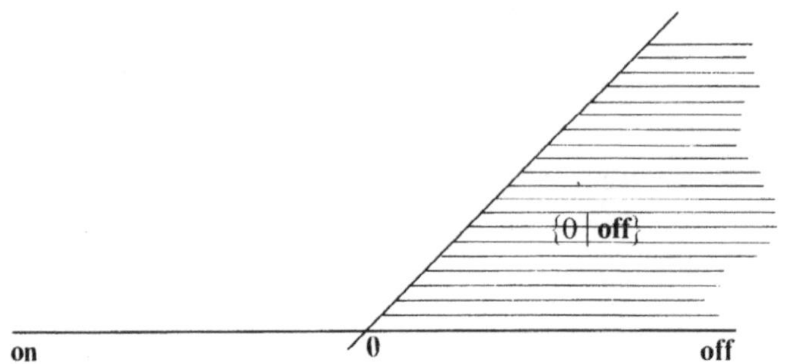

Figure 19. The Thermograph of **oof**.

We already know that **off** is the most negative game of all, and the thermograph of **oof** = 0|**off** (Fig. 19) shows it to be less than all positive numbers, but confused with all the others. We know that **tiny** is the smallest positive game of all. What about

$$\mathbf{ace},\ \mathbf{deuce},\ \mathbf{trey}, \ldots ?$$

It turns out that

$$\mathbf{ace} + \mathbf{ace} = \mathbf{deuce}, \qquad \mathbf{ace} + \mathbf{deuce} = \mathbf{trey}, \ldots$$

so that after the first three terms you only need to know about sums of **aces**.

But when we include their negatives as well, upsums begin to differ from downsums, for example

$$\begin{aligned} \mathbf{ace} \mathbin{\overset{\uparrow}{+}} (-\mathbf{ace}) &= \mathbf{joker}, \\ \mathbf{ace} \mathbin{\underset{\downarrow}{+}} (-\mathbf{ace}) &= -\mathbf{joker}, \end{aligned}$$

where **joker** and −**joker** are thoroughly different games:

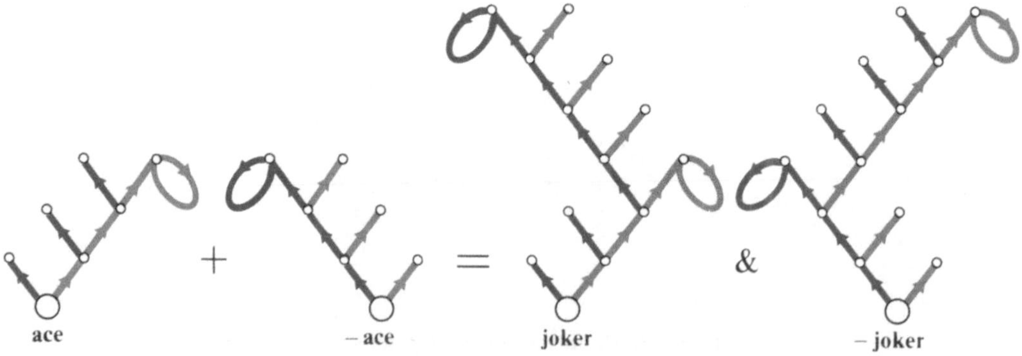

ace + – ace = joker & – joker

Now **ace** has atomic weight 1, so by repeatedly combining **ace**s and their negatives we shall get various games whose atomic weights are all whole numbers; for example, the game

deuce =	$0\vert\mathbf{ace}$	has atomic weight 2 and is called $2\clubsuit$, and
	$0\vert\mathbf{joker}$	has atomic weight 1 and is called $1\heartsuit$.

It turns out that for most atomic weights n, we get just 4 such games

$$n\clubsuit < n\diamondsuit < n\heartsuit < n\spadesuit.$$

But for atomic weights -1, 0,1 we also get

$$-\mathbf{ace} < -\mathbf{joker} < 0 < \mathbf{joker} < \mathbf{ace}$$

and their sums with **tiny** and **miny**.

Figure 20 shows you how these games are defined; Fig. 21 shows how they're compared and Fig. 22 how to add them. We use the notation

$$X+ = X + \mathbf{tiny}, X- = X + \mathbf{miny}.$$

$\mathbf{A} = \mathbf{ace} = 0\vert\mathbf{tiny}$	$\mathbf{A}- = \mathbf{on}\vert\mathbf{A}\vert\vert 0$	$\mathbf{A}+ = \mathbf{A}$
$\mathbf{J} = \mathbf{joker} = 0\vert\bar{\mathbf{A}}+$	$\mathbf{J}- = \mathbf{on}\vert\bar{\mathbf{J}}\vert\vert\bar{\mathbf{A}}$	$\mathbf{J}+ = \mathbf{J}$
$\bar{\mathbf{J}} = -\mathbf{joker} = \mathbf{A}-\vert 0$	$\bar{\mathbf{J}}+ = \mathbf{A}\vert\vert\mathbf{J}\vert\mathbf{off}$	$\bar{\mathbf{J}}- = \bar{\mathbf{J}}$
$\bar{\mathbf{A}} = -\mathbf{ace} = \mathbf{miny}\vert 0$	$\bar{\mathbf{A}}+ = 0\vert\vert\bar{\mathbf{A}}\vert\mathbf{off}$	$\bar{\mathbf{A}}- = \bar{\mathbf{A}}$

n	$n\clubsuit$	$n\diamondsuit$	$n\heartsuit$	$n\spadesuit$
...	...	...	...	...
$\bar{3} = -3$	$\bar{2}\clubsuit\vert 0$	$\bar{2}\diamondsuit\vert 0$	$\bar{2}\heartsuit\vert 0$	$\bar{2}\spadesuit\vert 0$
$\bar{2} = -2$	$\bar{1}\clubsuit\vert 0$	$\bar{1}\diamondsuit\vert 0$	$\bar{1}\heartsuit\vert 0$	$\mathbf{A}\vert 0$
$\bar{1} = -1$	$\clubsuit\vert 0$	$\mathbf{J}\vert 0$	$\heartsuit\vert 0$	$0\vert 2\spadesuit$
0	$1\clubsuit\vert 0$	$\mathbf{A}\vert\bar{1}\diamondsuit$	$1\heartsuit\vert\bar{\mathbf{A}}$	$0\vert\bar{1}\clubsuit$
1	$2\clubsuit\vert 0$	$0\vert 1\diamondsuit$	$0\vert\mathbf{J}$	$0\vert\spadesuit$
2	$0\vert\mathbf{A}$	$0\vert 1\diamondsuit$	$0\vert 1\heartsuit$	$0\vert 1\spadesuit$
3	$0\vert 2\clubsuit$	$0\vert 2\diamondsuit$	$0\vert 2\heartsuit$	$0\vert 2\spadesuit$
...	...	...	...	...
	$\clubsuit = 0\clubsuit$	$\diamondsuit = 0\diamondsuit$	$\heartsuit = 0\heartsuit$	$\spadesuit = 0\spadesuit$

For all games X in this table $X+ = X- = X$

deuce, trey, ... are alternative names for $2\clubsuit, 3\clubsuit, \ldots$ (but **ace** $\neq \clubsuit$).

Figure 20. Laying Out the Cards.

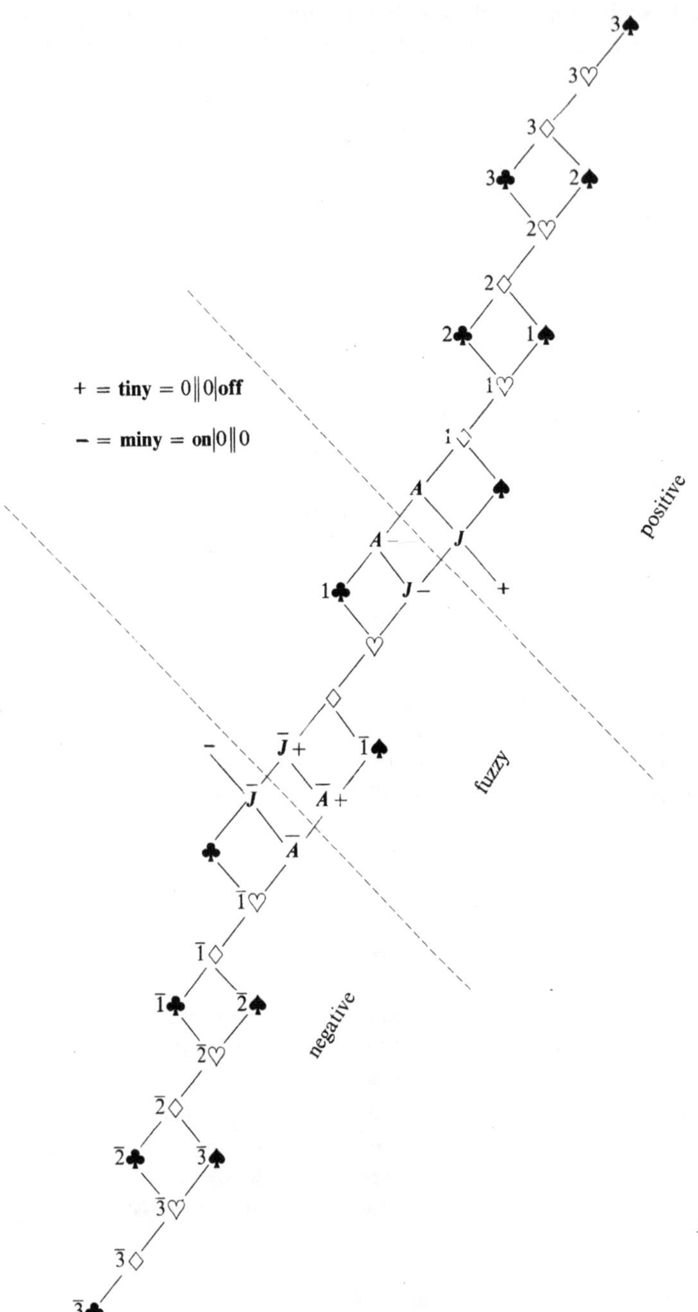

Figure 21. Putting Your Cards in Order.

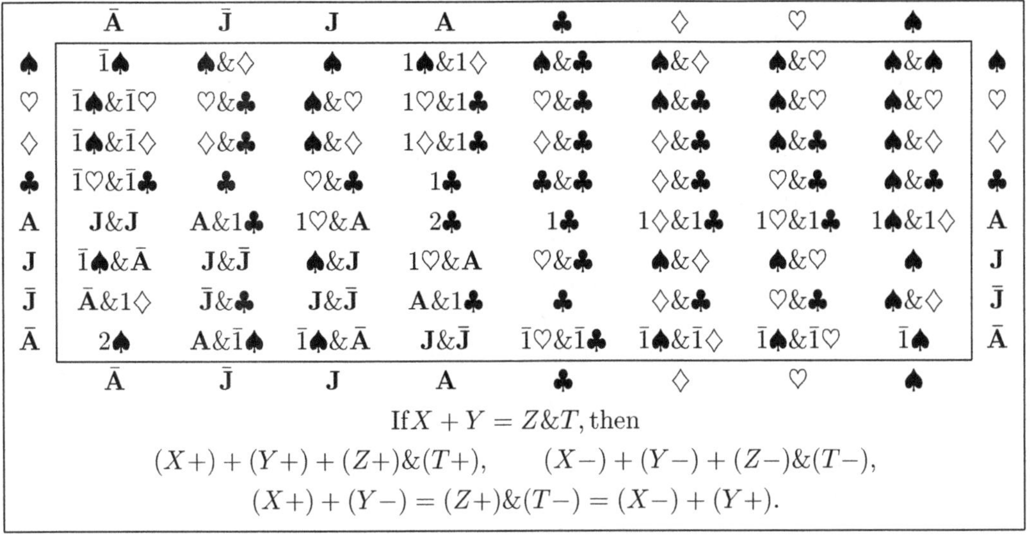

	$\bar{A}$	$\bar{J}$	J	A	$\clubsuit$	$\diamondsuit$	$\heartsuit$	$\spadesuit$	
$\spadesuit$	$\bar{1}\spadesuit$	$\spadesuit\&\diamondsuit$	$\spadesuit$	$1\spadesuit\&1\diamondsuit$	$\spadesuit\&\clubsuit$	$\spadesuit\&\diamondsuit$	$\spadesuit\&\heartsuit$	$\spadesuit\&\spadesuit$	$\spadesuit$
$\heartsuit$	$\bar{1}\spadesuit\&\bar{1}\heartsuit$	$\heartsuit\&\clubsuit$	$\spadesuit\&\heartsuit$	$1\heartsuit\&1\clubsuit$	$\heartsuit\&\clubsuit$	$\spadesuit\&\clubsuit$	$\spadesuit\&\heartsuit$	$\spadesuit\&\heartsuit$	$\heartsuit$
$\diamondsuit$	$\bar{1}\spadesuit\&\bar{1}\diamondsuit$	$\diamondsuit\&\clubsuit$	$\spadesuit\&\diamondsuit$	$1\diamondsuit\&1\clubsuit$	$\diamondsuit\&\clubsuit$	$\diamondsuit\&\clubsuit$	$\spadesuit\&\clubsuit$	$\spadesuit\&\diamondsuit$	$\diamondsuit$
$\clubsuit$	$\bar{1}\heartsuit\&\bar{1}\clubsuit$	$\clubsuit$	$\heartsuit\&\clubsuit$	$1\clubsuit$	$\clubsuit\&\clubsuit$	$\diamondsuit\&\clubsuit$	$\heartsuit\&\clubsuit$	$\spadesuit\&\clubsuit$	$\clubsuit$
A	$J\&J$	$A\&1\clubsuit$	$1\heartsuit\&A$	$2\clubsuit$	$1\clubsuit$	$1\diamondsuit\&1\clubsuit$	$1\heartsuit\&1\clubsuit$	$1\spadesuit\&1\diamondsuit$	A
J	$\bar{1}\spadesuit\&\bar{A}$	$J\&\bar{J}$	$\spadesuit\&J$	$1\heartsuit\&A$	$\heartsuit\&\clubsuit$	$\spadesuit\&\diamondsuit$	$\spadesuit\&\heartsuit$	$\spadesuit$	J
$\bar{J}$	$\bar{A}\&1\diamondsuit$	$\bar{J}\&\clubsuit$	$J\&\bar{J}$	$A\&1\clubsuit$	$\clubsuit$	$\diamondsuit\&\clubsuit$	$\heartsuit\&\clubsuit$	$\spadesuit\&\diamondsuit$	$\bar{J}$
$\bar{A}$	$2\spadesuit$	$A\&\bar{1}\spadesuit$	$\bar{1}\spadesuit\&\bar{A}$	$J\&\bar{J}$	$\bar{1}\heartsuit\&\bar{1}\clubsuit$	$\bar{1}\spadesuit\&\bar{1}\diamondsuit$	$\bar{1}\spadesuit\&\bar{1}\heartsuit$	$\bar{1}\spadesuit$	$\bar{A}$
	$\bar{A}$	$\bar{J}$	J	A	$\clubsuit$	$\diamondsuit$	$\heartsuit$	$\spadesuit$	

If $X+Y=Z\&T$, then

$$(X+)+(Y+)+(Z+)\&(T+), \qquad (X-)+(Y-)+(Z-)\&(T-),$$

$$(X+)+(Y-)=(Z+)\&(T-)=(X-)+(Y+).$$

Figure 22. Adding Up Your Cards.

The Degree of Loopiness

Of course every ordinary ender G has

$$G+(-G)=0.$$

But for loopy games γ we have

$$\gamma+(-\gamma)=d\&(-d),$$

where d is called the **degree of loopiness**, or just the **degree**, γ°, of γ. In symbols

$$\gamma^\circ=\gamma+(-\gamma),$$

the upsum of γ and its negative. The a,b,c Property implies that γ° measures the downsum absorbancy of γ:

$$\gamma+x\leq\gamma \text{ just if } x\leq\gamma^\circ,$$

and in particular,

$$\boxed{\gamma+x=\gamma \text{ for } 0\leq x\leq\gamma^\circ.}$$

THE DOWNSUM ABSORBANCY RULE

Our more mathematical readers will deduce from the properties of upsums and downsums that:

$$(-\gamma)^\circ=\gamma^\circ=\gamma^\circ+\gamma^\circ=(\gamma^\circ)^\circ,$$

$$\gamma^\circ + \delta^\circ \leq (\gamma + \delta)^\circ \leq \gamma^\circ + \delta^\circ, \qquad \gamma^\circ + \delta^\circ \leq (\gamma + \delta)^\circ \leq \gamma^\circ + \delta^\circ,$$

and similarly for more complicated sums,

$$\ldots \gamma^\circ + \delta^\circ + \ldots + \eta^\circ \ldots \leq (\ldots \gamma + \delta \ldots + \eta \ldots)^\circ \leq \ldots \gamma^\circ + \delta^\circ + \ldots + \eta^\circ \ldots .$$

A most important property is that

the degree of loopiness, γ°, is zero if γ is equal to some ender, and is otherwise strictly positive

So for genuinely loopy γ we have

$$\gamma^\circ \geq \textbf{tiny},$$

and therefore

$$\gamma + \textbf{tiny} = \gamma.$$

Lots of calculations are made easier using these ideas. Let's find the degree of **tiny**:

$$\textbf{tiny}^\circ = \textbf{tiny} + (-\textbf{tiny}).$$

Because we've added a negative quantity, this is $\leq$ **tiny**, but it must also be $\geq$ **tiny**, since **tiny** is a genuinely loopy game:

$$\textbf{tiny}^\circ = \textbf{tiny}.$$

We can now find

$$\textbf{tiny} + \textbf{tiny},$$

for since **tiny** has only red plums, **tiny** + **tiny** is a ripe plum tree and so already a stopper:

$$\textbf{tiny} + \textbf{tiny} = \textbf{tiny} + \textbf{tiny}.$$

But by the Downsum Absorbancy Rule

$$\textbf{tiny} + \textbf{tiny} = \textbf{tiny},$$

and so finally,

$$\textbf{tiny} + \textbf{tiny} = \textbf{tiny}.$$

Classes and Varieties

In the House of Cards the degrees of the various positions are

0	(for 0 only),
tiny	(for **tiny** and **miny**),
joker	(for $\mathbf{A}, \mathbf{A}-, \mathbf{J}, \mathbf{J}-, \bar{\mathbf{J}}, \bar{\mathbf{J}}+, \bar{\mathbf{A}}, \bar{\mathbf{A}}+$),
♠	(for $n♣, n♢, n♡, n♠; n = \ldots -2, -1, 0, 1, 2, \ldots.$).

The positions whose degree is ♠ (unlike the others) show a striking regularity—they form themselves naturally into *classes*, one for each n, and each class exhibits all four *varieties*:

♣, ♢, ♡, ♠.

This seems to happen more generally for all the games of any given **stable** degree d, i.e. one that satisfies

$$d \mathbin{\overset{\uparrow}{+}} d = d,$$

and in this case we will write $a\langle b\rangle$ for the game of *class* a and *variety* b. Formally, the **class** of a contains all the games of degree d between

$$a\langle -d\rangle = a \mathbin{\overset{\downarrow}{+}} (-d) \qquad \text{and} \qquad a\langle d\rangle = a \mathbin{\overset{\uparrow}{+}} d,$$

which are the smallest and largest members of the class. Although for loopy games

$$a \mathbin{\overset{\downarrow}{+}} \bar{a} < 0$$

we conjecture that

$$a\langle d\rangle \mathbin{\overset{\downarrow}{+}} \bar{a}\langle d\rangle \geq 0 \quad \text{(The Stability Condition)}$$

whenever a has degree d and d is stable. We say that a is **stable to the degree** d when this condition holds. It can be proved that sums of such games enjoy the same property.

The **variety** of a is the game

$$0\langle a\rangle = a \mathbin{\overset{\downarrow}{+}} \bar{a}\langle d\rangle = a \mathbin{\overset{\uparrow}{+}} \bar{a}\langle -d\rangle .$$

In the House of Cards, for example,

$$0\langle 3♡\rangle = ♡,$$

while the *class* of 3♡ contains all four cards

3♣, 3♢, 3♡, 3♠.

When a and b are stable to the degree d, there is a unique game

$$a\langle b\rangle = a\langle d\rangle \mathbin{\overset{\downarrow}{+}} 0\langle b\rangle = a\langle -d\rangle \mathbin{\overset{\uparrow}{+}} 0\langle b\rangle$$

which is in the same class as a, and has the same variety as b.

To add games given like this you just add the classes and add the varieties:

$$\begin{aligned} a\langle b\rangle \mathbin{\overset{\uparrow}{+}} c\langle d\rangle &= (a + c)\langle b \mathbin{\overset{\uparrow}{+}} d\rangle , \\ a\langle b\rangle \mathbin{\overset{\downarrow}{+}} c\langle d\rangle &= (a + c)\langle b \mathbin{\overset{\downarrow}{+}} d\rangle , \end{aligned}$$

We've written $a + c$ here because it doesn't matter whether you use $a \mathbin{\overset{\uparrow}{+}} c$ or $a \mathbin{\overset{\downarrow}{+}} c$.

There is a slight extension of this theory to cover games whose degree is less than d, provided they still satisfy the Stability Condition. Such a game has both

an **upsum variety**, $\quad a \mathbin{\overset{\uparrow}{+}} \bar{a}\langle -d\rangle$,

and

a **downsum variety**, $\quad a \mathbin{\overset{\downarrow}{+}} \bar{a}\langle d\rangle$;

you use the former when taking upsums, the latter for downsums. For example, for the game **ace**, the two varieties are

♣ and ♢,

and so **ace** can be replaced by 1♣ when taking upsums, and by 1♢ when taking downsums.

Exercise Two. Find the degree of **upon** $= \{\text{pass}|*\}$, and investigate all sums of **upon** and its negative.

No Highway

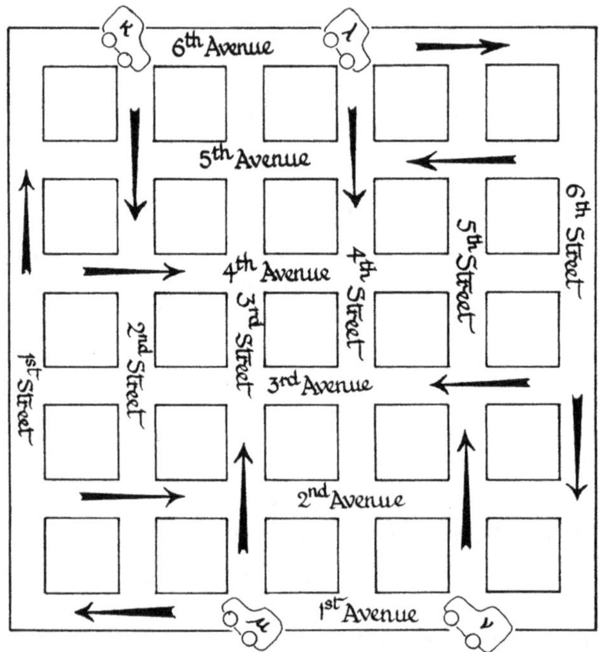

Figure 23. The Curious Cabs of Abnormal, Ill.

Figure 23 is a map of the town of Abnormal, Ill., in which you can see that all the streets and avenues are one-way. A hundred years ago this was a bustling town with tram-cars in every (N–S) street, operated by members of the Linksman's Labor Local; while the (E–W) avenues were served by elegant coaches manned by drivers of the Road Riders Regional. Sadly all this passed with the advent of the internal combustion engine, and only the two unions remain. For some time there has been an anxious truce; every cab (from the Greek Letter Cab Co.; you can see their new cab at the corner of 1st Avenue and 5th Street) has to carry one driver from each union, who takes the wheel when the cab is travelling in the appropriate direction. As there are only four cabs, competition for trips is fierce and there will always be a passenger wanting to travel one block in any legal direction (which is all the unions will allow!). The L.L.L. wants to keep the cabs running and will not allow two successive R.R.R. journeys. The R.R.R., on the other hand, are trying to wreck the system by bringing the cabs to a halt and are prepared to forego their right to make a trip on just 3 occasions.

In other words, LLL and RRR move alternately just one block in an appropriate direction, except that the RRR may have 3 pass moves. The RRR will win if the LLL cannot make a trip on their turn and the LLL win just by holding on for ever. What does this mean in loopy game terms?

Since all infinite plays win for the LLL, and

$$\gamma(\mathbf{on})^+ = \gamma^+,$$

we need only consider the onside of each position γ. So we want the onside of the sum

$$\kappa + \lambda + \mu + \nu - 3,$$

corresponding to the initial position and the 3 pass moves for the RRR.

We'll use the sidling process, starting with **on** at each intersection (Fig. 24(a)) and choosing a sensible order (Fig. 24(b)) in which to revise our estimates. We see almost immediately that the numbers in Fig. 24(c) are upper bounds, and just one more revision gives the final answers (Fig. 24(d)).

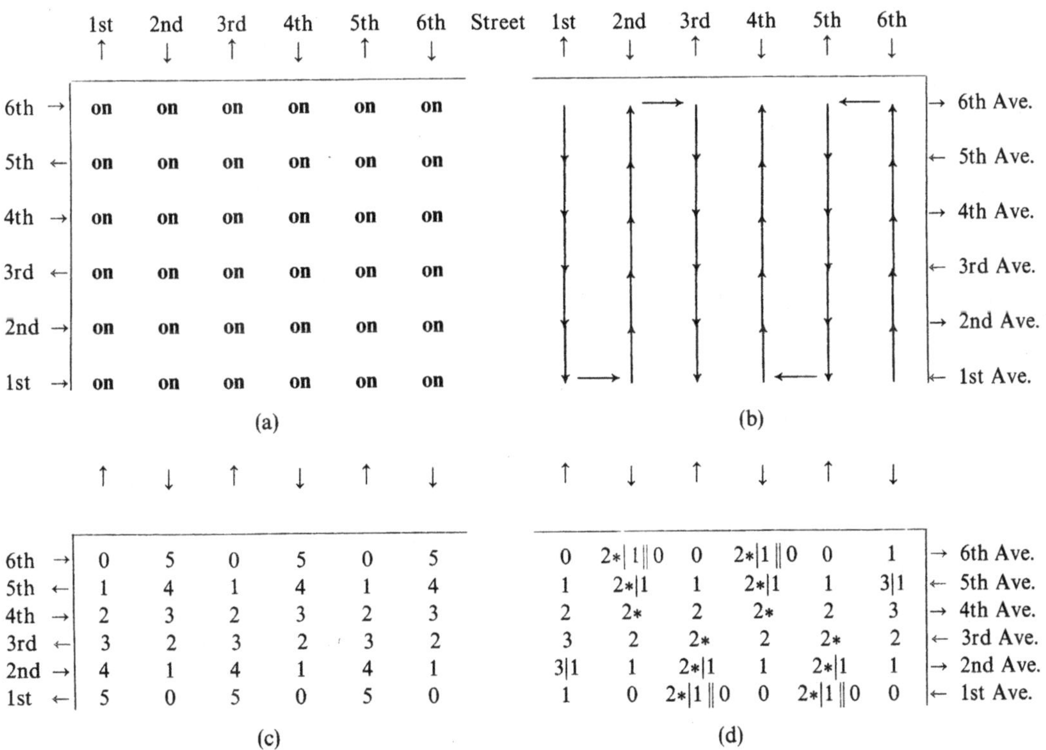

Figure 24. Sidling Grabs Idling Cabs.

<table>
<tr><th></th><th>1st St.
↑</th><th>2nd St.
↓</th><th>3rd St.
↑</th><th>4th St.
↓</th><th>5th St.
↑</th></tr>
<tr><td>10th Ave. →</td><td>0</td><td>$4{*}|3||2|||1||||0$</td><td>0</td><td>$4{*}|3||2|||1||||0$</td><td>0</td></tr>
<tr><td>9th Ave. ←</td><td>1</td><td>$4{*}|3||2|||1$</td><td>1</td><td>$4{*}|3||2|||1$</td><td>1</td></tr>
<tr><td>8th Ave. →</td><td>2</td><td>$4{*}|3||2$</td><td>2</td><td>$4{*}|3||2$</td><td>2</td></tr>
<tr><td>7th Ave. ←</td><td>3</td><td>$4{*}|3$</td><td>3</td><td>$4{*}|3$</td><td>3</td></tr>
<tr><td>6th Ave. →</td><td>4</td><td>$4{*}$</td><td>4</td><td>$4{*}$</td><td>4</td></tr>
<tr><td>5th Ave. ←</td><td>5</td><td>4</td><td>$4{*}$</td><td>4</td><td>$4{*}$</td></tr>
<tr><td>4th Ave. →</td><td>$5|3$</td><td>3</td><td>$4{*}|3$</td><td>3</td><td>4</td></tr>
<tr><td>3rd Ave. ←</td><td>3</td><td>2</td><td>$4{*}|3||2$</td><td>2</td><td>$4|2$</td></tr>
<tr><td>2nd Ave. →</td><td>$3|1$</td><td>1</td><td>$4{*}|3||2|||1$</td><td>1</td><td>2</td></tr>
<tr><td>1st Ave. ←</td><td>1</td><td>0</td><td>$4{*}|3||2|||1||||0$</td><td>0</td><td>$2|0$</td></tr>
</table>

Figure 25. No Highway in a Town of Five Streets and Ten Avenues.

It's just as easy to solve the same kind of game in any rectangular town. Figure 25 shows the values for a 5-street, 10-avenue town, and you can guess that the value

$$9*|8||7|||6||||5|||||4||||||3|||||||2||||||||1|||||||||0$$

will arise in a 20-avenue town. Where have we seen such things before?

In Chapter 6 we saw that such *overheated* values arise in Domineering and in Seating Boys and Girls (though sometimes they were disturbed by infinitesimals) and that they are best expressed as integrals. We write $\int x$ for $\int_{1*}^{1} x$ and $*\int_{1*}^{1} x$ for $* + \int x$, and find

$$\int 0 = 0, \qquad \int 1 = 1*, \qquad *\int \tfrac{1}{2} = 1*|0, \qquad *\int \tfrac{3}{4} = 2*|1||0, \ldots,$$

so that our previous display has value

$$*\int \tfrac{511}{512}.$$

We also have $\int * = 1|-1$, so $\int 1* = 2*|*$, etc., allowing us to rewrite Figs. 24(d) and 25 as Figs. 26 and 27.

$\int 0$	$*\int\frac{3}{4}$	$\int 0$	$*\int\frac{3}{4}$	$\int 0$	$*\int 1$
$*\int 1$	$\int 1\frac{1}{2}$	$*\int 1$	$\int 1\frac{1}{2}$	$*\int 1$	$\int 2*$
$\int 2$	$*\int 2$	$\int 2$	$*\int 2$	$\int 2$	$*\int 3$
$*\int 3$	$\int 2$	$*\int 2$	$\int 2$	$*\int 2$	$\int 2$
$\int 2*$	$*\int 1$	$\int 1\frac{1}{2}$	$*\int 1$	$\int 1\frac{1}{2}$	$*\int 1$
$*\int 1$	$\int 0$	$*\int\frac{3}{4}$	$\int 0$	$*\int\frac{3}{4}$	$\int 0$

Figure 26. Overheated Cabs.

$\int 0$	$*\int\frac{15}{16}$	$\int 0$	$*\int\frac{15}{16}$	$\int 0$
$*\int 1$	$\int 1\frac{7}{8}$	$*\int 1$	$\int 1\frac{7}{8}$	$*\int 1$
$\int 2$	$*\int 2\frac{3}{4}$	$\int 2$	$*\int 2\frac{3}{4}$	$\int 2$
$*\int 3$	$\int 3\frac{1}{2}$	$*\int 3$	$\int 3\frac{1}{2}$	$*\int 3$
$\int 4$	$*\int 4$	$\int 4$	$*\int 4$	$\int 4$
$*\int 5$	$\int 4$	$*\int 4$	$\int 4$	$*\int 4$
$\int 4*$	$*\int 3$	$\int 3\frac{1}{2}$	$*\int 3$	$\int 4$
$*\int 3$	$\int 2$	$*\int 2\frac{3}{4}$	$\int 2$	$*\int 3$
$\int 2*$	$*\int 1$	$\int 1\frac{7}{8}$	$*\int 1$	$\int 2$
$*\int 1$	$\int 0$	$*\int\frac{15}{16}$	$\int 0$	$*\int 1$

Figure 27. More Overheated Cabs.

Which union should win in Fig. 23? The four cabs are all at intersections of value $* \int \frac{3}{4}$ and so have total value

$$* + * + * + * + \int \left(\tfrac{3}{4} + \tfrac{3}{4} + \tfrac{3}{4} + \tfrac{3}{4}\right) = 0 + \int 3 = 3*,$$

and the allowance of 3 pass moves to the RRR subtracts 3 from this, leaving $*$, so whichever union starts first should win the game.

In a well-fought game the first four moves should move all four cabs, after which

$$* \int \tfrac{3}{4} + * \int \tfrac{3}{4} + * \int \tfrac{3}{4} + * \int \tfrac{3}{4}$$

will become

$$1 + * \int \tfrac{1}{2} + 0 + 1 + * \int \tfrac{1}{2} + 0.$$

The next two moves should both be with the cabs that are now on 2nd or 5th Avenue, to make the sum

$$2 * +1,$$

and the best next move is with the cab on 3rd or 4th Avenue. The cab-sum will then be

$$3,$$

making the whole game have value

$$3 - 3 = 0.$$

If RRR started, it's now LLL's turn and RRR can win by passing 3 times in succession. If LLL started, the RRR will find that their 3 passes aren't quite enough; whether they use them or not, LLL will always find a place to move.

Backsliding Toads-and-Frogs

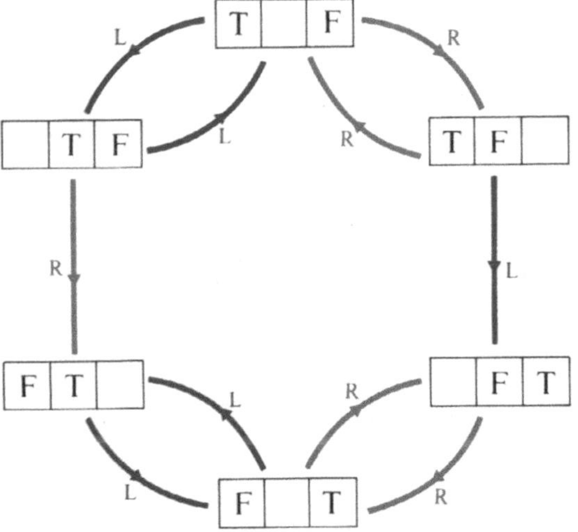

Figure 28. Backsliding Toads-and-Frogs.

This is played like ordinary Toads-and-Frogs (Chapter 1) except that the animals can *slide* backwards as well as forwards, but may not *jump* backwards. Here is the $(1,1)_1$ game (Fig. 28). Find the onsides, starting from $\Box FT \leq |\mathbf{on} = 0$, and check that the offsides are the same, with the values of the positions $0, *, 0, *, \ldots$ alternately; so whoever starts loses.

Try analyzing the $(1,1)_2$ (one toad, one frog, two places between) and $(2,2)_1$ games and check your results in the Extras.

Extras

Bach's Carousel

We tried quite hard to prove that every loopy game had stoppers for its onside and offside before Clive Bach, to whom some of the theory in this chapter is due, eventually found a game we call the Carousel (Fig. 29) which has several disturbing features:

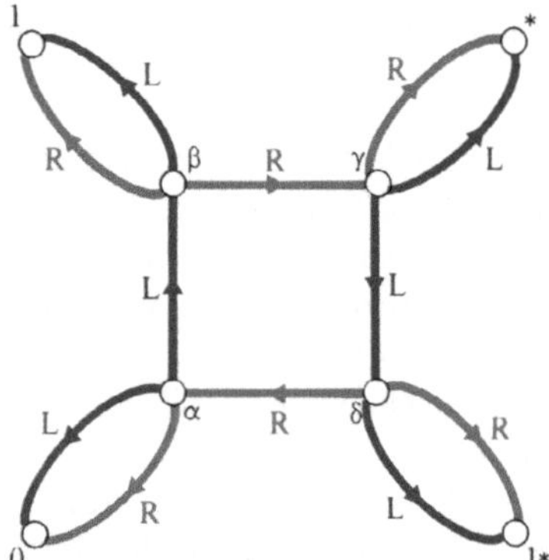

Figure 29. Bach's Carousel.

(i) its onsides and offsides are *not* equivalent to stoppers.

(ii) there is some dominated move that cannot be omitted,

(iii) if, in $\alpha - \alpha$, the first player always moves round the carousel, the second player cannot afford to do anything but make the mirror image move in the other component.

Statements (i) and (ii) are proved as follows. As regards the onsides α^+, β^+, γ^+, δ^+, the Carousel is equivalent to

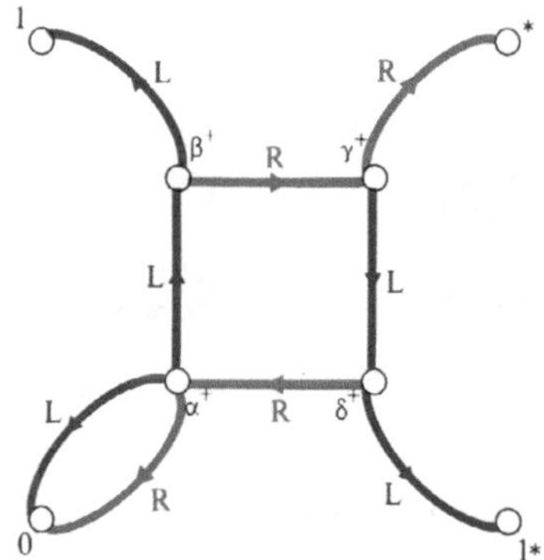

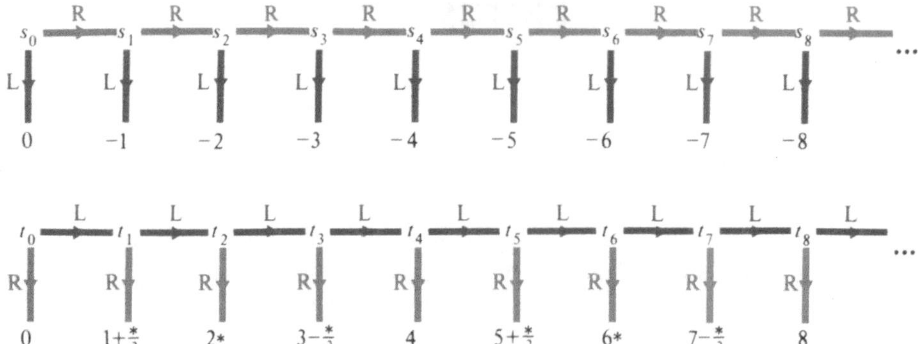

Figure 30. The Sides of the Sum of these Stoppers Are Not Equivalent to Stoppers.

obtained by omitting the dominated Left option $*$ of γ and Right's options 1 of β, $1*$ of δ. This has no *reversible* options from any position and only one dominated option (0 as a Left option of α). The proof of the Simplest Form Theorem now shows that any game in simplest form equivalent to α^+ must have the structure obtained by omitting this last dominated move.

But if we omit the dominated Left option 0 of α^+, the resulting game $\hat{\alpha}$, say, is not equivalent to α^+! You can check this by showing that Right can win $\hat{\alpha} - \alpha^+$. So α^+ has no simplest form, and cannot be a stopper.

The reason why 0 cannot be omitted, even though it's dominated, is roughly that Left can always arrive at a better position by taking another trip round the Carousel than by stepping off to 0 now. But there *are* circumstances in which he can win by stepping off sometime and *won't* win by going round and round for ever!

Question: Is there an alternative notion of simplest form that works for *all* finite loopy games (in particular, for the Carousel)?

Bach also points out that the sum of the two stoppers s_0 and t_0 in Fig. 30 apparently doesn't have an onside or an offside equivalent to a stopper. (In Fig. 30, "semi-star", $\frac{*}{2}$, is $\{*, \uparrow \mid \downarrow *, 0\}$.)

Getting **on** in Checkers

Here are two positions in the popular board game of checkers:

Left's black king is the only piece on the 2×2 board. Right has no possible moves, but Left can move to and fro. Formally, we have

$$\textbf{onto} = \{\textbf{onfro} \mid \}$$
$$\textbf{onfro} = \{\textbf{onto} \mid \}$$

Both **onto** and **onfro** are instances of the game called **on**. In ordinary checkers **dud** arises when the board contains only two kings, of opposite colors, located in opposite double corners.

Proof of the Sidling Theorem

If the sidling process for a game α^+ converges when started from **on**, it produces upper bounds

$$\alpha(\mathbf{on}), \qquad \beta(\mathbf{on}), \qquad \ldots$$

or say,

$$[\alpha], \qquad [\beta], \qquad \ldots$$

for the various positions

$$\alpha^+, \qquad \beta^+, \qquad \ldots$$

of α^+, that satisfy

$$[\alpha] \le \{[\alpha^L] \mid [\alpha^R]\}, \qquad [\beta] \le \{[\beta^L] \mid [\beta^R]\}, \qquad \ldots$$

(in fact with equality). We show that for *any* games $[\alpha], [\beta], \ldots$ satisfying these inequalities, we have

$$[\alpha] \le \alpha^+, \qquad [\beta] \le \beta^+, \qquad \ldots$$

(and if $[\alpha], [\beta], \ldots$ are also upper bounds we again have equality).

We suppose therefore that Left is given survival strategies in each of the games

$$\{[\alpha^L] \mid [\alpha^R]\} - [\alpha], \qquad \{[\beta^L] \mid [\beta^R]\} - [\beta], \qquad \ldots$$

and desires one in

$$\alpha^+ - [\alpha].$$

Since the new strategy is quite hard to find, we shall suppose that Right kindly places a potential infinity of his more mathematically inclined servants, messrs.

rado,	radon,	...	rademacher,	...
r_0,	r_1,	...	r_m,	...

at the disposal of Left, and allows him to use the Great Hall, and various furnishings, of *The Wright House*, which is rather a grand establishment.

On the far table in Fig. 31 is set up the real game $\alpha^+ - [\alpha]$ which Left is to play against his real opponent, Right. But, even before play starts. Left instructs r_0 to bring in an additional table on which is set up the difference game

$$\{[\alpha^L] \mid [\alpha^R]\} - \{[\alpha^L] \mid [\alpha^R]\} = \alpha_0 - \alpha_0, \text{ say,}$$

and a seat labelled S_0, to be placed near the games

$$\{[\alpha^L] \mid [\alpha^R]\} = \alpha_0 \text{ and } -[\alpha].$$

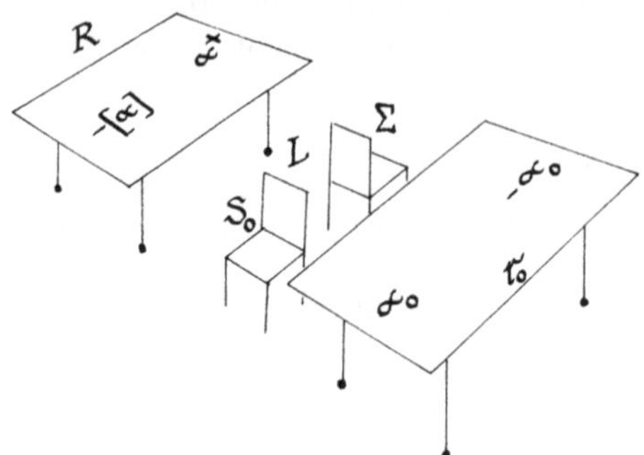

Figure 31. Sidling Strategy—Starting Seats.

Left has, by the hypotheses of the theorem, a survival strategy, which we also call S_0, in the sum of these two games. The seat marked Σ, which was already in the Hall, is placed near the games

$$\alpha^+ \text{ and } -\alpha_0.$$

As the game proceeds, Left occasionally instructs a new footman, r_i, to bring in a new seat, S_i, and a new table on which is set up a position of the form $\alpha_i - \alpha_i$. Footman r_i is detailed from then on to respond to a Left move in either α_i or $-\alpha_i$ with the mirror-image move in the other. In Fig. 32 we see a number of these tables, all marked with the positions in which they were originally set up.

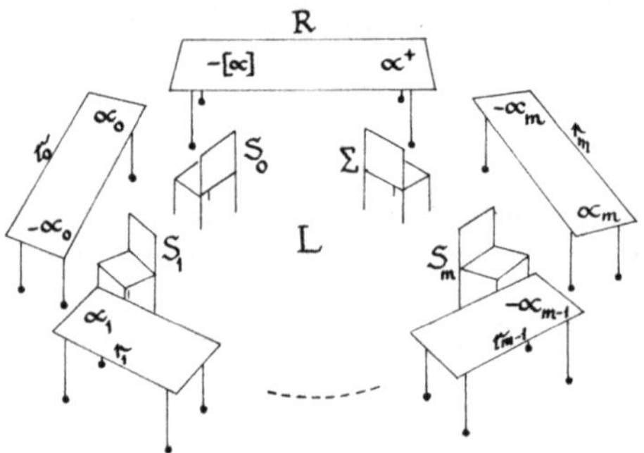

Figure 32. Right, Rado, Radon, ..., Round to Rademacher.

The seats $S_0, S_1, \ldots, S_m$ and Σ are placed *between* adjacent tables, and each corresponds to a strategy, of sorts, for playing the two games nearest to it. The strategies S_i are easiest to describe. When seat S_i was first brought in, the games it was put next to were in a position of the form

$$\{[\beta^L] \mid [\beta^R]\} - [\beta]$$

for some position β of α; strategy S_i is Left's survival strategy for this game given by our hypotheses.

The two games nearest to seat Σ will usually have the form

$$\beta = \{\beta^L|\beta^R\} \text{ and } -\{[\beta^L] \mid [\beta^R]\}$$

for some position β of α. The "strategy" Σ is then the following sequence of actions. If Right, or rademacher, makes a move in either of these games, Left is to make the corresponding move in the other, making the compound position have the form

$$\gamma - [\gamma]$$

for some position $\gamma = \beta^L$ or β^R of α. He then instructs a new footman, r_{m+1}, to bring in a new table on which is set up the difference game

$$\{[\gamma^L] \mid [\gamma^R]\} - \{[\gamma^L] \mid [\gamma^R]\},$$

and a new seat, S_{m+1}, to be placed near to the games

$$\{[\gamma^L] \mid [\gamma^R] \text{ and } -[\gamma]$$

for whose sum he has a survival strategy of the same name, S_{m+1}. The seat Σ is then repositioned next to

$$\gamma = \{\gamma^L|\gamma^R\} \text{ and } -\{[\gamma^L] \mid [\gamma^R]\}.$$

Left's total strategy is therefore this. To any move, whether played by his real opponent, Right, or one of the footmen $r_0, r_1, \ldots$, he replies with the response given by the strategy corresponding to the nearest seat. The strategies S_i are those for various differences $\alpha_i - \alpha_{i-1}$ with $\alpha_i \geq \alpha_{i-1}$ that are given us by the hypotheses, while the strategy Σ requires just one "imitation" move, and a call for a new seat and table to be inserted. It is plain that this compound strategy always gives Left a reply in a game somewhere in the Hall, but it's not entirely clear that he will eventually respond to any move in the real game with another move in that game, and perhaps even less clear that he avoids loss in that game if it continues indefinitely. We now proceed to establish these facts.

If infinitely many tables are brought in, then Left certainly makes infinitely many moves against his opponent in α^+, and since all infinite plays in α^+ are wins for Left, he avoids loss.

If not, we can imagine that no more tables are brought in than those shown in Fig. 32, say, and therefore

$$\text{sign}\left(\alpha^+\right) = \text{sign}\left(\alpha_m\right) = 0.$$

Now strategies $S_0, \ldots, S_m$ arrange that

$$\text{sign}([\alpha]) \leq \text{sign}(\alpha_0) \leq \text{sign}(\alpha_1) \leq \ldots \leq \text{sign}\left(\alpha_m\right) = 0,$$

and so $\text{sign}([\alpha]) = 0$ or $-$. If $\text{sign}([\alpha]) = -$, then Left certainly makes infinitely many moves in the real game $\alpha^+ - [\alpha]$ and satisfies the REMAIN,ONTOP survival conditions for comparing loopy games. If $sign([\alpha]) = 0$, the total play in all components is finite and Left makes the last move. This cannot be against a manservant, who always replies, so must be against his real opponent, Right. The Sidling Theorem is therefore proved.

A form of the sidling process appears in Bob Li's paper which first inspired us to attempt a general theory of loopy games. The main result of that paper is essentially the Sidling Theorem in the case when all onsides and offsides are numbers.

Answer to Exercise One

It's easy to check there's no infinite alternating sequence of moves from any position of this game, so α, β, γ, δ are stoppers and we need only sidle in from one side, say from **off**.

From the defining equations

$$\alpha = \{\delta|\}, \qquad \beta = \{\alpha|\alpha\}, \qquad \gamma = \{|\alpha\}, \qquad \delta = \{\gamma|\beta\},$$

we find successively

$$\alpha \geq \{\mathbf{off}|\} = 0,$$
$$\beta \geq \{0|0\} = *,$$
$$\gamma \geq \{|0\} = -1,$$
$$\delta \geq \{-1|*\} = 0,$$
$$\alpha \geq \{0|\} = 1,$$
$$\beta \geq \{1|1\} = 1*,$$
$$\gamma \geq \{|1\} = 0,$$
$$\delta \geq \{0|1*\} = 1,$$
$$\alpha \geq \{1|\} = 2,$$
$$\beta \geq \{2|2\} = 2*,$$
$$\gamma \geq \{|2\} = 0,$$
$$\delta \geq \{0|2*\} = 1,$$

and the sidling has finished. In fact, sidling in from **on** is quicker.

tis and **tisn**

The onsides of

$$\mathbf{tis} = \{\mathbf{tisn}|\} \text{ and } \quad \mathbf{tisn} = \{|\mathbf{tis}\}$$

can be approximated by

$$\mathbf{tisn} \leq \{|\mathbf{on}\} = 0,$$
$$\mathbf{tis} \leq \{0|\} = 1,$$
$$\mathbf{tisn} \leq \{|1\} = 0$$

and the offsides by

$$\mathbf{tis} \geq \{\mathbf{off}|\} = 0,$$
$$\mathbf{tisn} \geq \{|0) = -1,$$
$$\mathbf{tis} \geq \{-1|\} = 0,$$

so

$$\mathbf{tis} = 1 \,\&\, 0, \qquad \mathbf{tisn} = 0 \,\&\, -1.$$

To answer the question asked earlier about the outcome of

$$\mathbf{tis} + *$$

when all infinite plays count as draws, note that Left wins by moving in $*$, since Right has no move in **tis**. But if Right starts he must play to $\mathbf{tis} + 0$ and play will now be infinite, so the game is counted a draw.

upon

The game **upon** has the stable degree

$$d = \uparrow^{\mathbf{on}} = \{0 \mid -\mathbf{upon}*\} \text{ ("up-onth").}$$

In this case by adding **upon**s and their negatives we get one class for each $n = ...-2, -1, 0, 1, 2, ...$ and each class contains just two varieties

$$(\mathbf{upon} \times n)\langle -d \rangle \text{ and } (\mathbf{upon} \times n)\langle d \rangle .$$

So to add two such things you just add the n's the way you did at school and the d's like this:

$$d + d = d, \ (-d) + (-d) = -d,$$

$$d + (-d) = d\&(-d) = (-d) + d.$$

Backsliding Toads-and-Frogs

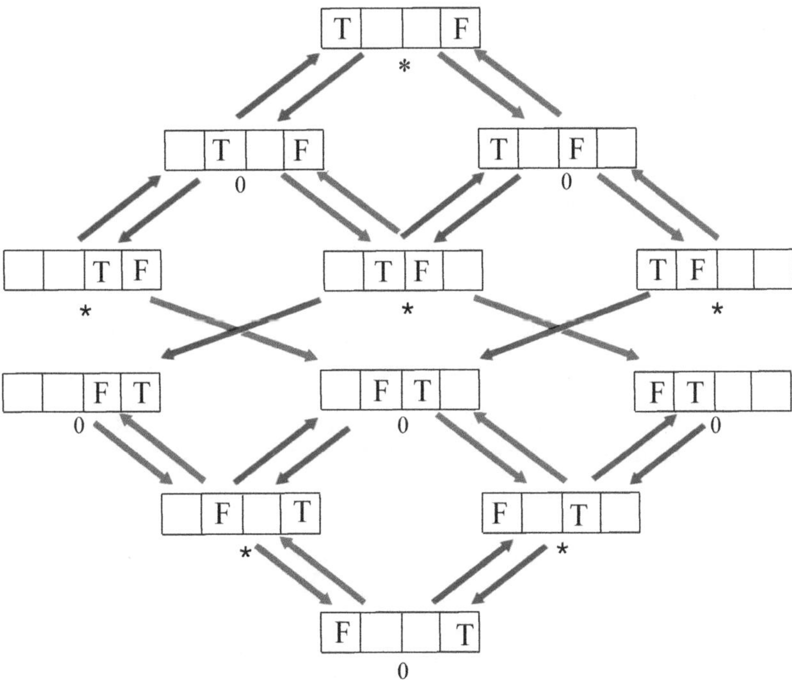

Figure 33. Latent Loopiness in Backsliding $(1, 1)_2$ Toads-and-Frogs.

After sidling in on $(1,1)_2$ Toads-and-Frogs we obtain only the not very loopy looking values 0 and $*$ as in Fig. 33. However there is latent loopiness which may show itself in actual play.

□FFTT ⟷(R) F□FFTT ⟷(R) FF□TT ⟷(L) FFT□T ⟷(L) FFTT□

off **off** **hot** **on** **on**

F□TFT →(R) **on**; FTF□T →(L) **off**

□FTFT ⟷(R) F□TFT ⟷(L) FT□FT ⟷(R) FTF□T ⟷(L) FTFT□

0 * 0 * 0

□TFFT →(R) 0; TF□FT →(L) 0

□TFFT ⟷(L) T□FFT ⟷(R) TF□FT ⟷(R) TFF□T ⟷(L) TFFT□

* 0 * 1|* 1

FT□TF →(R) 0; FTTF□ →(L) 0

□FTTF ⟷(R) F□TTF ⟷(L) FT□TF ⟷(L) FTT□F ⟷(R) FTTF□

−1 *|−1 * 0 *

□TFTF →(R) *; TF□TF →(L) −1; TF□TF →(R) 1; TFTF□ →(L) *

□TFTF ⟷(L) T□FTF ⟷(R) TF□TF ⟷(L) TFT□F ⟷(R) TFTF□

0 * 0 * 0

T□TFF →(R) *; TTF□F →(L) *

□TTFF ⟷(L) T□TFF ⟷(L) TT□FF ⟷(R) TTF□F ⟷(R) TTFF□

1 1|* * *|−1 −1

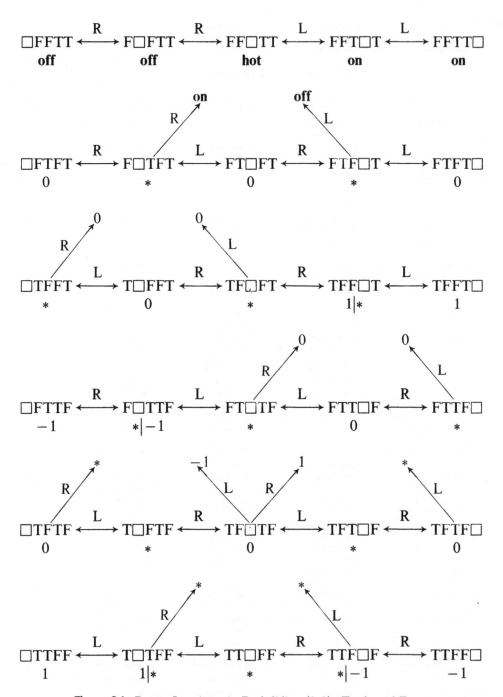

Figure 34. Patent Loopiness in Backsliding $(2,2)_1$ Toads-and-Frogs.

The sidling process for the $(2,2)_1$ game is best done in the six separate stages of Fig. 34, since then you use the fact that any jumping move takes you to an already analyzed position. You'll see that patently loopy values arise in the innermost portion (only).

KOs in Go

Loopy positions called "KOs" arise in the classical Asian board game known as Go. However, all dialects of the rules of Go forbid a player to make any move which repeats the same position that occurred on the board after his prior turn. Nevertheless, Go endgame positions typically decompose into sums of several regions which can be effectively treated as if they were independent games. A region containing a ko is then locally loopy, because the rule banning immediate recapture of the ko is global.

Thermographs, masts, means, and temperatures now all play important roles in the study of Go engames. Berlekamp, Mueller, and Spight analyzed a collection of dozens of ko positions in 1996 which occurred in professional games. In 2002 Spight pioneered the study of a more refined methodology called "dogmatism." Dogmatism always yields precise results, although the conditions of its applicability to real professional endgames can sometimes seem somewhat arbitrary. Spight and Fraser also report on interesting results on so-called "super-KOs," which are Go positions whose game graphs contain loops of lengths ≥ 4. A sidling-like technique, which obtains results via the rapid convergence of a small number of successive approximations, was introduced and analyzed by Fraser.

References and Further Reading

E. Berlekamp, M. Mueller, and Bill Spight, Generalized Thermography: Algoritms, Implementation, and application to Go endgames, Technical Report 96-030, ICSI Berkeley, 1996.

J. H. Conway, *On Numbers and Games*, A K Peters, Ltd., Natick, MA, 2001, Chapter 16.

J. H. Conway, Loopy games, in Béla Bollobás (ed.): Advances in Graph Theory (Cambridge Combinatorial Conference 1977), *Ann. Discrete Math.*, **3**(1978) 55–74.

James Alan Flanigan, An analysis of some take-away and loopy partizan graph games, Ph.D. dissertation, UCLA, 1979.

Bill Fraser, Computer-Assisted Thermographic Analysis of Go Endgames, Ph.D. dissertation, UC Berkley, 2002, also online at math.berkeley.edu/~bfraser/thesis.ps

Robert Li Shuo-Yen. Sums of Zuchswang games, *J. Combin. Theory, Ser. A*, **21**(1976) 52–67.

Ahiezer S. Shaki, Algebraic solutions of partizan games with cycles, *Math. Proc. Cambridge Philos. Soc.*, **85**(1979) 227–246.

Bill Spight, Extended Thermography for multiple KOs in go, 2002.

Bill Spight, Evaluating KOs in a Neutral Threat Environment: Preliminary Results, 2002.

-12-

Games Eternal – Games Entailed

If it was not for the entail I should not mind it.
Jane Austen, *Pride and Prejudice*, ch. 23.

The hornèd moon, with one bright star
Within the nether tip
Samuel Taylor Coleridge, *The Ancient Mariner*, pt. iii.

What happens when you play a sum of games in a way that breaks some of the usual moving and ending conditions? The two main theories of this chapter are about *impartial* games—C. A. B. Smith's for impartial loopy games, that might last forever, and our own new theory for games with entailing or complimenting moves. The harder theories of partizan loopy games and misère play of impartial ordinary ones are treated in the neighboring chapters.

Have you noticed how the almonds at parties tend to get left in the nut-bowl because they're too hard to crack? Here's a nutty little game you can use to befuddle the other guests.

Fair Shares and Varied Pairs

Figure 1. Anyone for Fair Shares and Varied Pairs?

Take up to 10 almonds and arrange them in heaps on the carpet—we suggest you start with lots of twos and ones. Then challenge the guest who interests you most to a game of **Fair Shares and Varied Pairs** in which the moves are:

Divide any heap into two or more *equal-sized* heaps, FAIR SHARES

or

Unite any *two* heaps of *different sizes.* VARIED PAIRS

You win by completely separating the almonds (into heaps of size 1) after which your opponent has no legal move. There's always at least one move from any other position, e.g. to **shatter** some heap into heaps of size 1.

The funny thing about this game is that you can often return to a position you've already seen, not always with the same player to move. In fact it can be quite hilarious to watch an accomplished player spin the game round and round in several different circles before sneaking in the win. When the winning move comes, the opponent's too giddy to work out just where they stepped off the carousel (especially if this came after another glass of wine).

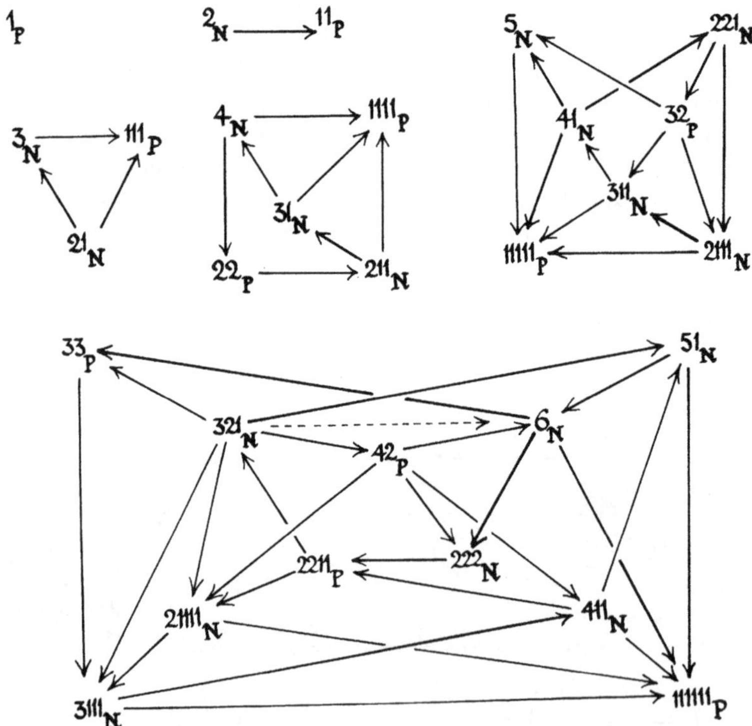

Figure 2. Fair Shares and Varied Pairs Positions with up to Six Almonds. (The dotted arrow refers to the variant Fair Shares and Unequal Partners.)

To keep your own balance in this game you'll obviously need to know the $\mathcal{P}$-positions. For instance, in the 9-almond game these are

72, 54, 5211, 4311, 4221, 333, 321111, 222111, 111111111

It's easy to find these on a graph of the game—mark a position $\mathcal{P}$ only if all its options have been marked $\mathcal{N}$ (in particular if it has no option) and mark a position $\mathcal{N}$ if it has an option already marked $\mathcal{P}$. We've done this in Fig. 2 for the games with up to 6 almonds.

To win, of course, you must always move to a $\mathcal{P}$-position. But if this is all you know, you might just conceivably find yourself going round and round Fig. 3 for the rest of time. You're always moving to $\mathcal{P}$-positions (in the boxes) but somehow you never seem to win. Why is this?

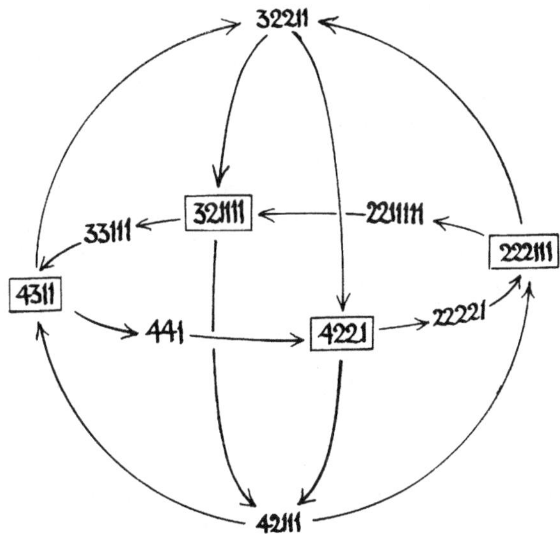

Figure 3. Round and Round the World with Nine Almonds.

How Soon Can You Win?

The other thing you need to know about is the *remoteness* of the positions. Recall from Chapter 9 that the remoteness is the number of moves in which the winner can force his win and recall that a position is $\mathcal{P}$ or $\mathcal{N}$ according as its remoteness is even or odd. But games with cycles can also have positions with infinite remoteness from which neither player can force a win because with best play the game continues for ever.

For such games we work out the remoteness of the positions in stages:

Stage 0. Assign remoteness 0 to all terminal positions.

Stage 1. Assign remoteness 1 to any position with an option of remoteness 0.

Stage 2. Assign remoteness 2 to any new position *all* of whose options have been assigned remoteness 1.

Stage 3. Assign remoteness 3 to any new position with an option of remoteness 2.

... ...

Stage $2n$. Assign remoteness $2n$ to those new positions *all* of whose options have already been assigned smaller odd numbers (the greatest of which will be $2n-1$).

Stage $2n+1$. Assign remoteness $2n+1$ to any new position which has an option of remoteness $2n$.

There May Be Open Positions ($\mathcal{O}$-Positions)

When you reach a stage where you're unable to assign any more remoteness according to these rules, all other positions have remoteness infinity and the game will continue for ever with the best play. We call these **open positions** or **$\mathcal{O}$-positions**. Figure 4 shows all the $\mathcal{P}$-positions (boxed) and some of the $\mathcal{N}$-positions in Fair Shares and Varied Pairs played with up to 10 almonds, arranged according to their remoteness. If you want to win, remember the $\mathcal{P}$-positions, especially the ones of low remoteness, but to avoid giving the show away you should not always take the shortest possible win. You can get a longer ride on the carousel by moving to $\mathcal{P}$-positions of higher remoteness than necessary for a few times. Against a novice it's often a good idea to move to $\mathcal{N}$-positions of high remoteness, and wait for the inevitable mistake.

It's a remarkable fact that with 10 almonds or less there are no infinite games with best play. As soon as we get to 11 almonds, the situation changes dramatically:

There is 1 $\mathcal{P}$-position $\boxed{11111111111}$ with remoteness 0,
10 $\mathcal{N}$-positions 2111111111, 311111111, 41111111, ... with remoteness 1, and
45 $\mathcal{O}$-positions (all those with 2 or more splittable heaps).

So the game is rather dull—even when played between an expert and a tyro who can only see one move ahead, it will go on for ever.

If n is the sum of 2 primes, p and q, then in the n-almond game the "Goldbach" position $\boxed{p,q}$ has remoteness 2, but we believe that for every $n > 10$ most positions are $\mathcal{O}$-positions.

In any game we expect there to be more $\mathcal{N}$-positions than $\mathcal{P}$-positions, because a position with a $\mathcal{P}$-option is automatically $\mathcal{N}$. If the game has a large number of cycles we should expect also a large number of $\mathcal{O}$-positions. (Fair Shares and Varied Pairs with 10 almonds or less was atypical.) So when analyzing such a game by hand it's best to concentrate on the $\mathcal{P}$-positions. You first find all the terminal positions—these are the $\mathcal{P}$-positions of remoteness 0. Next you attack the $\mathcal{P}$-positions of remoteness 2, which are those for which every move can be reversed to a terminal position. Continue until you have exhausted all the $\mathcal{P}$-positions (or yourself). It's usually a waste of ink to write down the $\mathcal{N}$- or $\mathcal{O}$-positions.

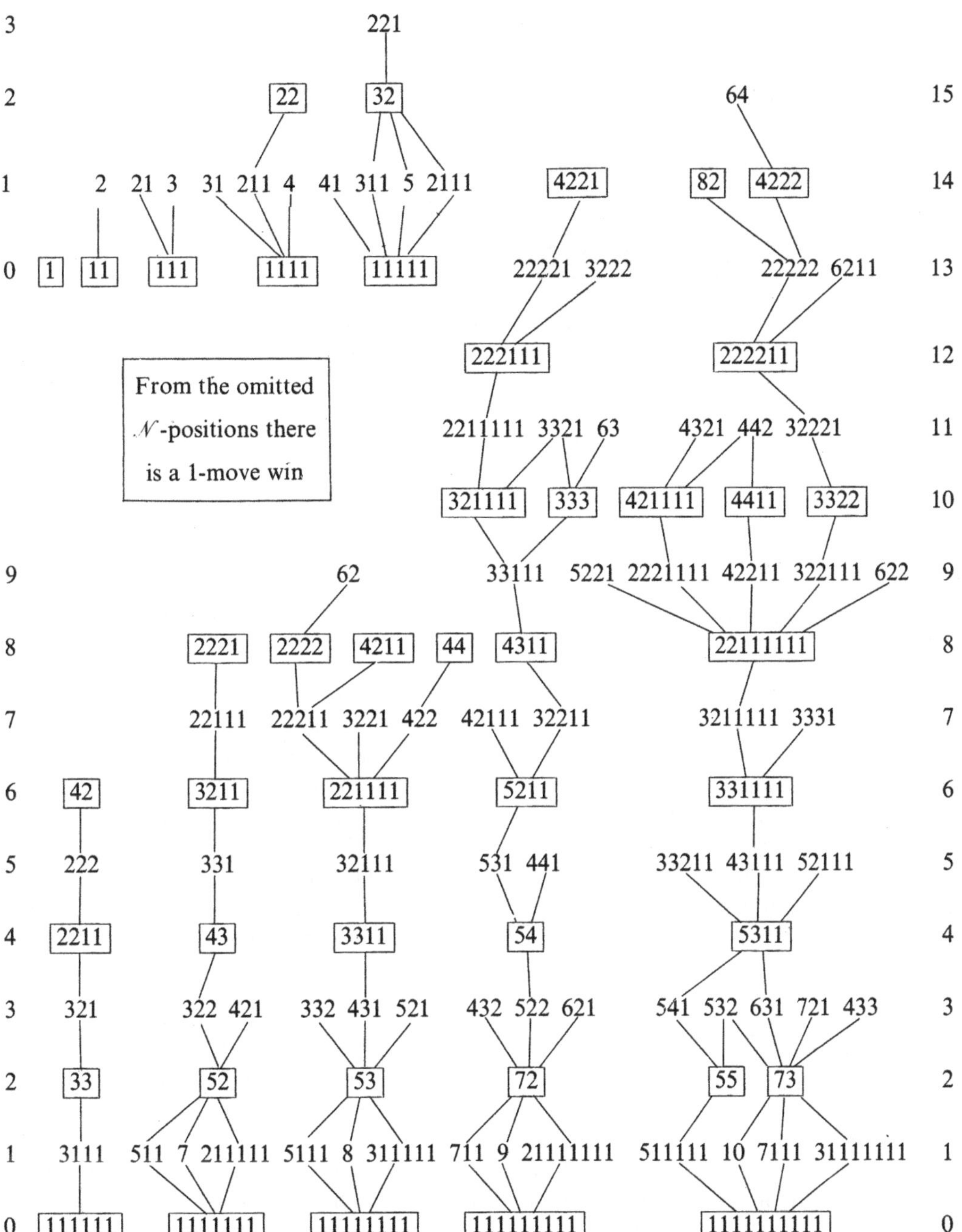

Figure 4. Positions with up to Ten Almonds.

De Bono's *L*-Game

In *The Five-day Course in Thinking* Edward de Bono introduces this little game (Fig. 5).

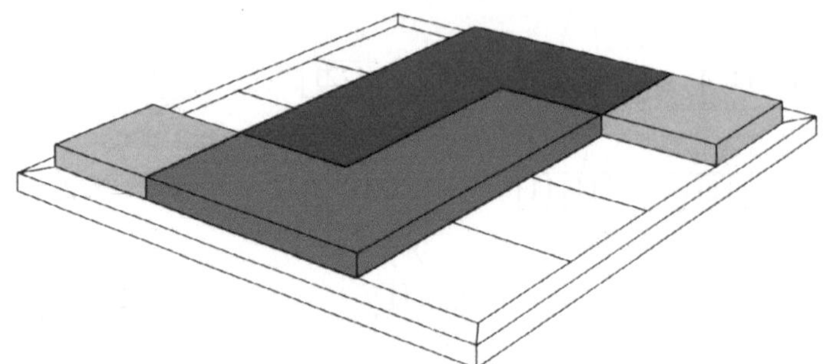

Figure 5. How to Start the *L*-Game.

It is played on a 4×4 board; each player has his own *L*-shaped piece which may be turned over, and there are 2 neutral 1×1 squares. A move has two parts:

> You *must* lift up your own *L*-piece and put it back on the board in *another* position.
> You *may*, if you wish, change the position of *one* of the two neutral pieces.

If you can't move, because there's only the one place for your *L*-piece, you lose.

To within symmetries of the board, there are 82 different positions for the *L*s, and therefore $62 \times 28 = 2296$ distinct positions in all. It's not hard to show by hand that there are just 15 $\mathcal{P}$-positions of remoteness 0; one could then look for those of remoteness 2, etc., so reducing the work to just 29 2-move backward analyses. Rather than thinking forwards or laterally, think backwards! Fortunately, however, we don't have to think at all, because V. W. Gijlswijk, G. A. P. Kindervater, G. J. van Tubergen & J. J. O. O. Wiegerink used a computer to find an optimal move from each of the 2296 positions and identified the 29 $\mathcal{P}$-positions, 1006 $\mathcal{N}$-positions and 1261 $\mathcal{O}$-positions. The $\mathcal{N}$- and $\mathcal{P}$-positions are distributed as follows:

remoteness	0	1	2	3	4	5	6	7	8	9	Total
$\mathcal{P}$-positions	15		3		3		5		3		29
$\mathcal{N}$-positions		768		27		81		11		119	1006

If you're capable of seeing just one move ahead, you don't need the complete list given in their paper. If you move to one of the 29 positions appearing in Fig. 6, and play well thereafter, you will *win*. Otherwise the game should be a *draw* by infinite repetition, unless your opponent has just left you in one of these positions with the roles reversed. In this case he probably knows what he is doing and you can expect to *lose*.

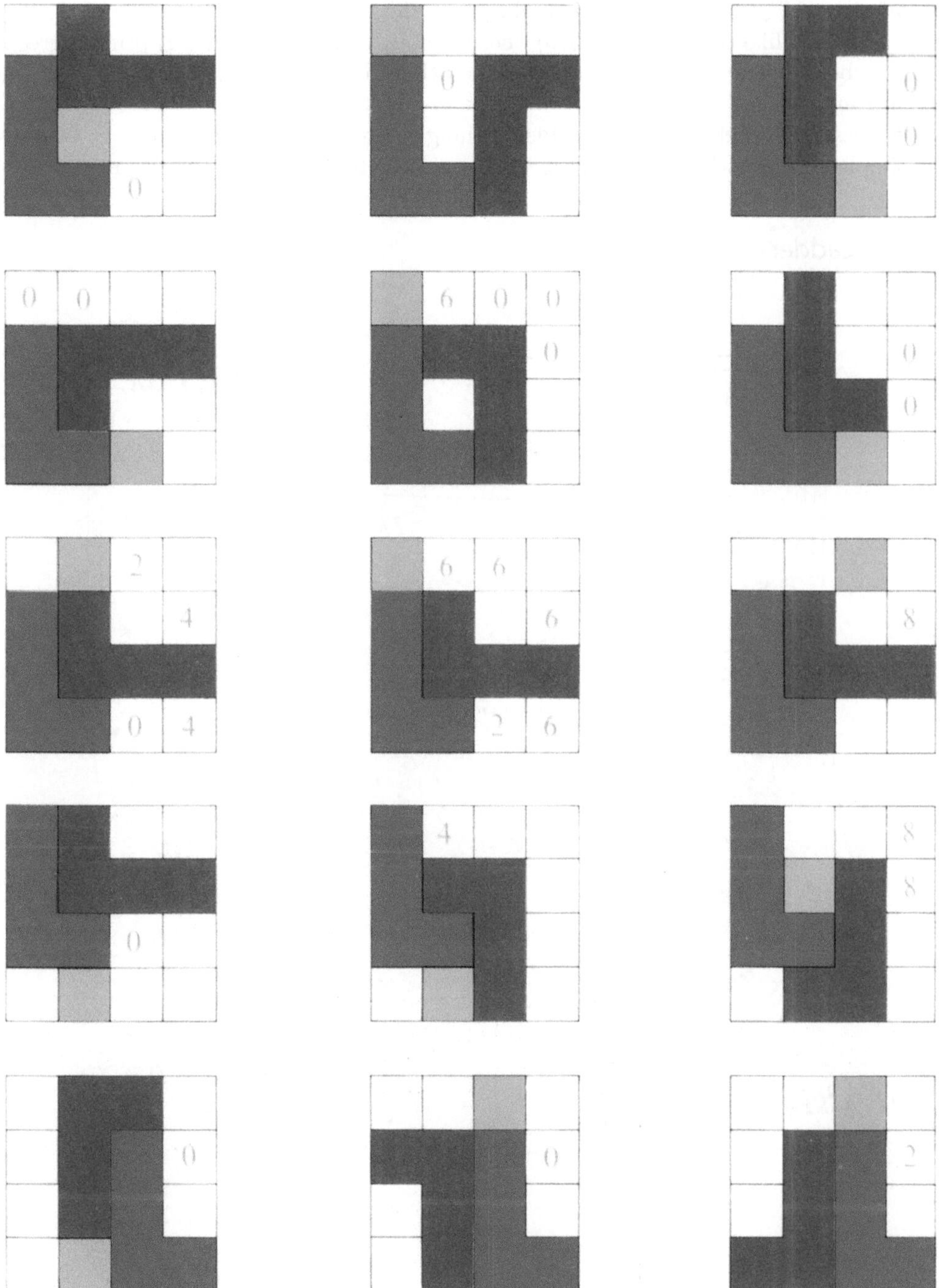

Figure 6. Remotenesses of the 29 $\mathcal{P}$-Positions of De Bono's L-Game.

The positions in Fig. 6 have been oriented so that the red (active) L looks like an L. Try to make *your* L look like the blue (inactive) one. One of the neutral squares is printed green; the green numbers indicate possible places for the other, and tell you the remoteness of the resulting position.

In the Extras you'll find a longest sensible finishing game (remoteness 9) and N. E. Goller's elegantly simple strategy which guarantees at least a draw from the initial position.

Adders-and-Ladders

Figure 7. A Cheap Game of Snakes-and-Ladders—Jimmy to Move.

Great-Aunt Maude gave Jimmy and Ginny a compendium of games this Christmas, but most of the games were cheaply produced and not too interesting. The Snakes-and-Ladders game, for example, was only a 5×5 card with a plastic 4-numbered die that didn't roll very well, and counters you couldn't distinguish between, as in Fig. 7. They soon found that by sliding the die along the table they could "roll" any number the liked from 1 to 4, and so they abandoned its use and just moved any counter they liked up to 4 squares onwards. They agreed that any number of counters could be on the same square and that the person who moved the last counter home was the winner.

Ginny soon noticed that when all the counters were on the top row it was what she called "that awful game with matches". In our language the game is a sum of games with one counter and the values of the top row are

$$\boxed{21} = *4, \quad \boxed{22} = *3, \quad \boxed{23} = *2, \quad \boxed{24} = *1, \quad \boxed{25} = 0.$$

When you land on $\boxed{20}$ you instantly climb the ladder to $\boxed{23}$, so

$$\boxed{20} = \boxed{23} = *2,$$

and similarly

$$\boxed{11} = \boxed{21} = *4, \quad \boxed{9} = \boxed{25} = 0, \quad \boxed{8} = \boxed{24} = *1$$

And because there's a snake from $\boxed{19}$ to $\boxed{7}$,

$$\boxed{19} = \boxed{7},$$

but we don't yet know the value of these two squares.

But Jimmy discovered that $\boxed{10} = \boxed{18}$ is a $\mathcal{P}$-position:

$$\boxed{10} = \boxed{18} = 0,$$

for after any Ginny move from $\boxed{18}$, Jimmy can get home in one:

	Ginny's move		Jimmy's reply	
$\boxed{18}$	$\longrightarrow$	$\boxed{19} = \boxed{7}$	$\longrightarrow$	$\boxed{9} = \boxed{25}$
$\boxed{18}$	$\longrightarrow$	$\boxed{20} = \boxed{23}$	$\longrightarrow$	$\boxed{25}$
$\boxed{18}$	$\longrightarrow$	$\boxed{21}$	$\longrightarrow$	$\boxed{25}$
$\boxed{18}$	$\longrightarrow$	$\boxed{22}$	$\longrightarrow$	$\boxed{25}$

Figure 8 shows the values we shall find for Adders-and-Ladders (and a simplified copy of the board for reference). A square at the foot of a ladder or head of a snake is not really a genuine position because a counter cannot stay there. Their values (red in Fig. 8) are simply copied from the other end of the (l)adder. The *loopy values* ∞_{012} and ∞_{12} will be explained later.

Because the game has cycles we are sometimes forced to compute the value of a position before we have evaluated all its options. Thus Jimmy was able to show that

$$\boxed{18} = 0,$$

despite the question mark in

$$\boxed{19} = ?, \quad \boxed{20} = *2, \quad \boxed{21} = *4, \quad \boxed{22} = *3,$$

because the *questionable option is reversible* to 0:

$$\boxed{19} = \boxed{7} \longrightarrow \boxed{9} = \boxed{25} = 0.$$

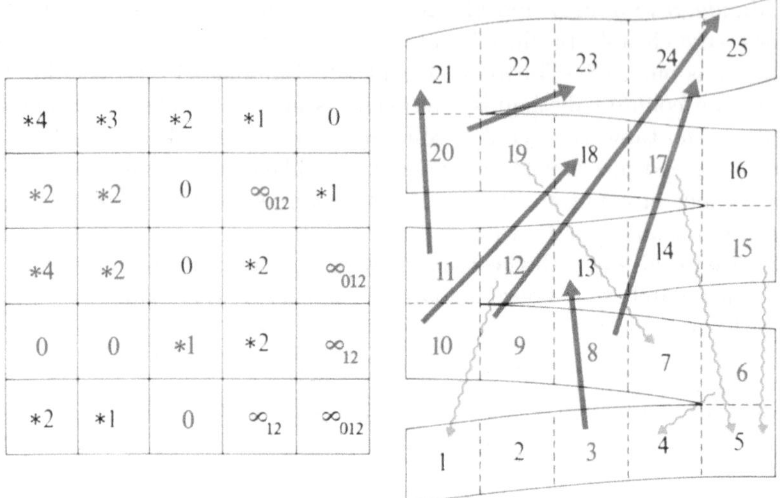

Figure 8. Adders-and-Ladders Values.

In general:

> If we have worked out the values
> $*a,\ *b,\ *c,\ \ldots$
> of *some* options of a position G and
> $m = \operatorname{mex}(a, b, c, \ldots)$
> then we can assert
> $G = *m$
> *provided that* the *other* options of G
> all have reversing moves to positions
> already known to have value $*m$.

CALCULATING ANSWERS BY SMITH'S RULE

When the proviso is satisfied G can be replaced by $*m$ in any sum for the usual reason—it has moves corresponding to all those of $*m$ and any other move is just a delaying move which can be reversed to $*m$.

Jimmy and Ginny didn't analyze their game much further, but *we* can, using this idea. For example the options of $\boxed{16}$ are:

$$\boxed{17} = ?, \quad \boxed{18} = 0, \quad \boxed{19} = ?, \quad \boxed{20} = *2$$

and $\operatorname{mex}(0, 2) = 1$. So, because $\boxed{17}$ and $\boxed{19}$ have reversing moves to $*1$:

$$\boxed{17} = \boxed{5} \rightarrow \boxed{8} = \boxed{24} = *1, \qquad \boxed{19} = \boxed{7} \rightarrow \boxed{8}$$

we can evaluate

$$\boxed{16} = *1.$$

Sometimes we can find the value of a position before evaluating *any* of its options. For example we could have proved that $\boxed{3} = \boxed{13} = 0$ by observing that *all* its options are reversible:

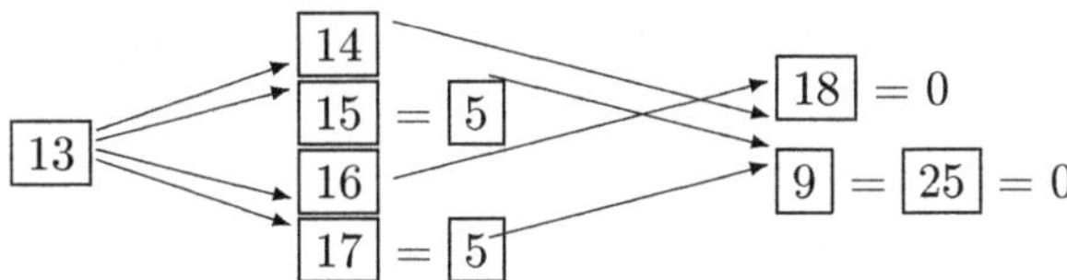

Here's how we justify the other non-loopy values in Fig. 8:

$\boxed{19} = \boxed{7} = *2$	$\boxed{14} = *2$	$\boxed{2} = *1$	$\boxed{12} = \boxed{1} = *2$
because			
$\boxed{8} = \boxed{24} = *1$	$\boxed{15} = \boxed{5} \rightarrow \boxed{7} = *2$	$\boxed{3} = \boxed{13} = 0$	$\boxed{2} = *1$
$\boxed{9} = \boxed{25} = 0$	$\boxed{16} = *1$	$\boxed{4} \rightarrow \boxed{8} = *1$	$\boxed{3} = \boxed{13} = 0$
$\boxed{10} = \boxed{18} = 0$	$\boxed{17} = \boxed{5} \rightarrow \boxed{7}$	$\boxed{5} \rightarrow \boxed{8}$	$\boxed{4} \rightarrow \boxed{7} = *2$
$\boxed{11} = \boxed{21} = *4$	$\boxed{18} = 0$	$\boxed{6} = \boxed{4}$	$\boxed{5} \rightarrow \boxed{7}$
and			
mex(0,1,4) = 2	mex(0,1) = 2	mex(0) = 1	mex(0,1) = 2

But now there's no other position we can work out this way. In general when you've forged ahead as far as you can with Smith's Rule there often comes a stage when the analysis falters because there's no other position to which the Rule applies. All positions that are still unlabelled at this stage are called **loopy** and are *not* equivalent to Nim-heaps.

> If the non-loopy options of a loopy position G have values
> $*a$, $*b$, $*c$, …,
> we say that G has value
> $\infty_{abc\ldots}$.

VALUES OF LOOPY POSITIONS

In particular,

$$\boxed{17} = \boxed{15} = \boxed{5} = \infty_{012}$$
$$\boxed{6} = \boxed{4} = \infty_{12}$$

as we see from their options:

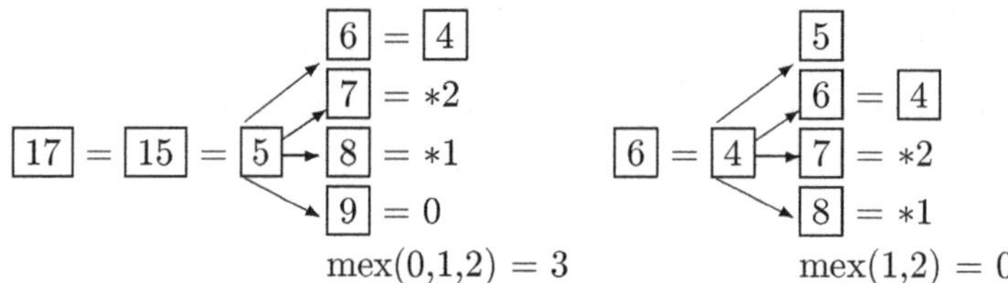

and the remarks:

$\boxed{5}$'s unmarked option $\boxed{4}$ has no move to an established $*3$;

$\boxed{4}$'s unmarked option $\boxed{4}$ has no move to an established 0

(although $\boxed{4}$'s other unmarked option, $\boxed{5}$, *can* be reversed to 0).

There are rules for adding these loopy games and Games with ordinary nimber values. The usual Nim Addition Rule allows us to replace the non-loopy Components by a single Nim-heap, $*n$, say.

With *no* loopy components,

$$*n \text{ is } \begin{cases} \mathcal{P} \text{ if } n = 0, \\ \mathcal{N} \text{ if not.} \end{cases}$$

With *one* loopy component,

$$\infty_{abc\ldots} + *n \text{ is } \begin{cases} \mathcal{N} \text{ if } n \text{ is one of } a, b, c, \ldots, \\ \mathcal{O} \text{ otherwise} \end{cases}$$

and has total value $\infty_{ABC\ldots}$

where $A = a \ddagger n$, $B = b \ddagger n$, $\ldots$.

With *more* loopy components,

$\infty_{abc\ldots} + \infty_{\alpha\beta\gamma\ldots} + \cdots$ is always $\mathcal{O}$,

and has total value ∞.

ADDING LOOPY VALUES

What should Jimmy do from the Adders-and-Ladders position of Fig. 7? The values of the three counters, read from Fig. 8, are

$$\boxed{1} = *2, \quad \boxed{5} = \infty_{012}, \quad \boxed{22} = *3,$$

so the position is really

$$\infty_{012} + *1 = \infty_{103}.$$

Jimmy's unique winning move is from $\boxed{5}$ to $\boxed{8}$ (and so up the ladder to $\boxed{24}$) since $\boxed{8} = *1$. If instead he moves from $\boxed{1}$ to $\boxed{3} = \boxed{13} = 0$, or from $\boxed{1}$ to $\boxed{4} = \infty_{12}$, the value becomes

$$\infty_{012} + *3 = \infty_{321} \text{ or } \infty_{012} + \infty_{12} + *3 = \infty$$

and Ginny will make sure the game continues forever (unless Jimmy makes another stupid move). Had he moved from $\boxed{1}$ to $\boxed{2} = *1$, the value would be

$$\infty_{012} + *2 (= \infty_{230})$$

and Ginny would instantly move from $\boxed{5}$ to $\boxed{7} = *2$.

Just How Loopy Can You Get?

Loopiness may be latent, patent or even blatant!

Some positions admit circular chains of moves that need never arise in best play, even when you're adding them to other (non-loopy) games. This kind of loopiness is really illusory; unless the winner wants to take you on a trip you won't notice it. Such positions have the same kind of values $*n$ as non-loopy ones, and are only **latently loopy**.

Other positions reveal their loopiness in their values:

$$\infty_{abc\ldots}$$

and so are **patently loopy**. However if one of the subscripts is zero the loopiness only affects best play when the game arises in a sum. If the game is played by itself it should take only finitely many moves.

If no subscript is zero the position is **blatantly loopy**. The best move from such a position is to another of the same type and so a well-played game will last for ever.

Fair Shares and Varied Pairs exemplifies all three cases. Table 1 will show that with 9 almonds or less, all the loopiness is latent, while with 11 or more almonds almost all positions are blatantly loopy, often having the loopiest possible value ∞. With exactly 10 almonds there are lots of loopy positions, but none of them is blatantly so. For example 4321 with value ∞_0 is an $\mathcal{N}$-position (move to 421111) but $4321 + *1$ is an $\mathcal{O}$-position.

Corrall Automotive Betterment Scheme

Figure 9 is a view of Corrall Island, which was a Charming Antipodean Beauty Spot before the Corrall Automotive Betterment Scheme, financed by the island's three wealthy landowners, built those expensive new super-highways. The island's two political parties (Left and Right) are now running for election and the gallop polls predict a walk-over for whichever party puts the last car on the scrap-heap. The parties have already enforced the one-way system indicated and each will alternately bribe a chauffeur to drive his car along one highway (in the proper direction) to the next intersection. You will see that there is no legal move away from the scrap-heap. How should we advise the ruling party, which must make the first move?

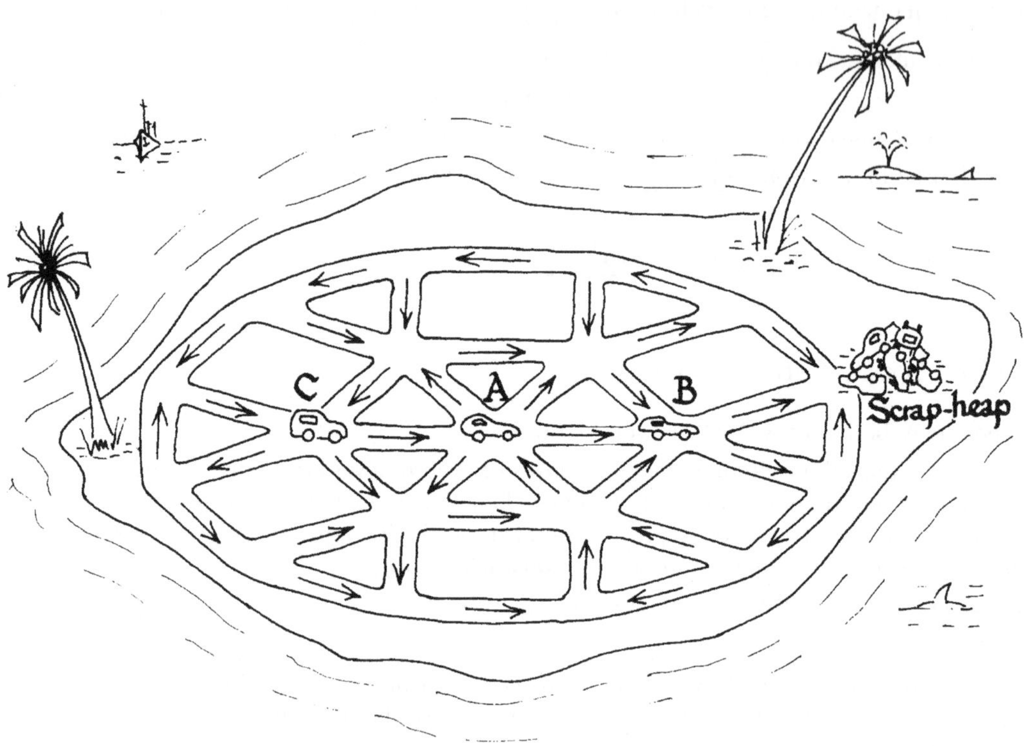

Figure 9. Corrall Automotive Betterment Scheme.

This is exactly the same kind of game as Adders-and-Ladders, but played on a rather more ingenious graph. Figure 10 shows the values, and with each non-loopy value, the **stage** at which we justified it, starting with the scrap-heap at stage 0. The C.A.B.S. Rule enables us to

> assign value $*m$ to a position G at stage k,
> *provided that* G has options which have been
> assigned values $*a, *b, *c, \ldots$ with $mex(a, b, c, \ldots) = m$ at earlier
> stages, *and that* all other options have moves to positions
> assigned $*m$ at stages earlier than k.

CALCULATING ASSIGNMENTS BY STAGES

For example, the node F which we labelled $*1$ at stage $5(*1_{@5})$ had an option E labelled 0 at stage 1, and the only other option (C, for which we later found the value ∞_{13}) had a reversing move to D, labelled $*1$ at stage 4. This is actually the unique winning move from Fig. 9.

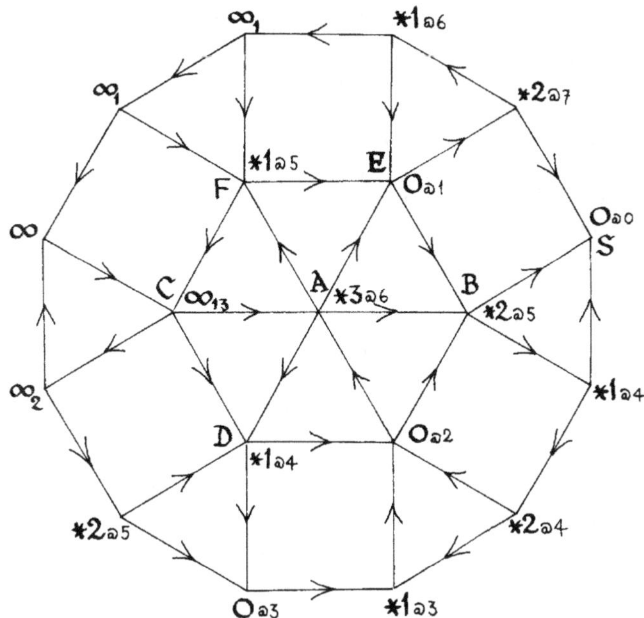

Figure 10. Graph, Values and Stages for Corrall Automotive Betterment Scheme.

So long as you label at least one node at each stage, and continue until Smith's Rule is no longer applicable, you'll get the correct values. But if you always label each node at the *earliest* stage you can (as we did for Corrall Island), you'll have a bonus:

> The number of the earliest stage at which you can label a position G with value $*n$ is half the remoteness of (G + a Nim-heap of size n).

CELERITY ADDS BONUS STEINHAUS

Sharing Out Other Kinds of Nut

You can use these ideas to play a sum of games of Fair Shares and Varied Pairs. This time you use some heaps of Cashews, some heaps of Almonds, some heaps of Brazils, and later on we'll add a heap of Sweets. No heap may contain two different kinds of nut, but otherwise the rules are as before—you may split a heap into *equal* smaller ones or unite two heaps which contain *different* numbers of the *same* kind of nut. Table 1 contains the value and the earliest stage at which it can be assigned for each position with at most 10 nuts of one kind. You might like to practise your skill by recalculating some of these from the graphs shown in Fig. 2.

1	$0_{@0}$	1111111	$0_{@0}$	111111111	$0_{@0}$	1111111111	$0_{@0}$
		52	$0_{@1}$	72	$0_{@1}$	55	$0_{@1}$
11	$0_{@0}$	7	$*1_{@1}$	54	$0_{@2}$	73	$0_{@1}$
2	$*1_{@1}$	43	$0_{@2}$	5211	$0_{@3}$	5311	$0_{@2}$
		511	$*1_{@2}$	4311	$0_{@4}$	331111	$0_{@3}$
111	$0_{@0}$	3211	$0_{@3}$	321111	$0_{@5}$	22111111	$0_{@4}$
3	$*1_{@1}$	31111	$*1_{@3}$	333	$0_{@5}$	3322	$0_{@5}$
21	$*2_{@2}$	2221	$0_{@4}$	222111	$0_{@6}$	421111	$0_{@5}$
		22111	$*1_{@4}$	9	$*1_{@6}$	4411	$0_{@5}$
1111	$0_{@0}$	211111	$*2_{@4}$	4221	$0_{@7}$	222211	$0_{@6}$
22	$0_{@1}$	331	$*2_{@4}$	711	$*1_{@7}$	4222	$0_{@7}$
4	$*1_{@2}$	421	$*1_{@5}$	51111	$*1_{@8}$	82	$0_{@7}$
211	$*1_{@3}$	322	$*2_{@5}$	3111111	$*1_{@9}$	22222	$*1_{@7}$
31	$*2_{@3}$	4111	$*2_{@5}$	441	$*1_{@9}$	10	$*2_{@8}$
		61	$*3_{@5}$	2211111	$*1_{@10}$	622	∞_{01}
11111	$0_{@0}$			42111	$*1_{@10}$	64	∞_{02}
32	$0_{@1}$	11111111	$0_{@0}$	21111111	$*2_{@10}$	91	∞_{02}
5	$*1_{@1}$	53	$0_{@1}$	33111	$*2_{@10}$		
311	$*1_{@2}$	3311	$0_{@2}$	3222	$*1_{@11}$	The 25 partitions	
221	$*1_{@3}$	221111	$0_{@3}$	3321	$*1_{@11}$	of 10 not listed	
2111	$*2_{@3}$	2222	$0_{@4}$	32211	$*2_{@11}$	here each have	
41	$*2_{@4}$	4211	$0_{@4}$	411111	$*2_{@11}$	value ∞_0.	
		44	$0_{@4}$	522	$*2_{@11}$		
111111	$0_{@0}$	8	$*1_{@5}$	6111	$*3_{@11}$		
33	$0_{@1}$	422	$*1_{@6}$	22221	$*2_{@12}$		
2211	$0_{@2}$	71	$*2_{@6}$	63	$*2_{@12}$		
42	$0_{@3}$	22211	$*1_{@7}$	432	$*3_{@13}$		
222	$*1_{@3}$	5111	$*1_{@8}$	81	$*3_{@13}$		
51	$*1_{@4}$	611	$*3_{@8}$	531	$*4_{@14}$		
6	$*2_{@4}$	311111	$*1_{@9}$	621	$*4_{@14}$		
3111	$*1_{@5}$	332	$*1_{@9}$				
411	$*2_{@5}$	41111	$*2_{@9}$				
21111	$*2_{@6}$	2111111	$*2_{@10}$				
321	$*3_{@7}$	3221	$*2_{@11}$				
		32111	$*3_{@11}$				
		521	$*3_{@11}$				
		62	$*4_{@11}$				
		431	$*3_{@12}$				

Table 1. Values and Stages for Fair Shares and Varied Pairs.

Fair Shares and Unequal Partners

In this variation you are also allowed to combine three or more heaps, all of different sizes (the dotted arrow in Fig. 2 represents such a move). This alters only a few of the values; you can find the details in the Extras.

Sweets and Nuts, and Maybe a Date?

But let's imagine that the guest you're interested in has another game in mind, and promises to make a date if you win the following variation. You play an ordinary game of Fair Shares and Varied Pairs with 9 almonds, but as an alternative to moving, either player may eat a number of the toffees you can also see in Fig. 1. What result should you expect if you're to start from the position of that figure?

The Additional Subtraction Games

Remember that in the Subtraction Game $S(a, b, c, ...)$ you are allowed to take away a or b or c or ... beans from any non-empty heap. What happens if some of $a, b, c, ...$ are negative? Since we then find ourselves adding beans to a heap, such a game might go on forever, and we need the Smith Theory to analyze it. Table 2 gives some typical values.

n	0	1	2	3	4	5	6	7	8	9	10	11	12	13	14	15
S(-1,2,5)	$0_{@0}$	$0_{@1}$	$2_{@3}$	$1_{@2}$	$0_{@1}$	$3_{@5}$	$2_{@4}$	$1_{@3}$	$0_{@2}$	$3_{@6}$	$2_{@5}$	$1_{@4}$	$0_{@3}$	$3_{@7}$	$2_{@6}$	$1_{@5}$
S(-2,4,5)	$0_{@0}$	$1_{@2}$	$0_{@1}$	$0_{@1}$	$1_{@3}$	$2_{@5}$	$3_{@7}$	$1_{@4}$	$2_{@6}$	∞_{12}	∞_{23}	∞_{13}	∞_{12}	∞_{2}	∞	∞
S(-1,3,8)	$0_{@0}$	$1_{@2}$	$0_{@1}$	$1_{@3}$	$0_{@2}$	$1_{@4}$	$0_{@3}$	$1_{@5}$	$2_{@7}$	$3_{@7}$	$2_{@6}$	$0_{@3}$	$1_{@5}$	$0_{@4}$	$1_{@6}$	$0_{@5}$
S(-2,3,8)	$0_{@0}$	$0_{@1}$	$0_{@2}$	$2_{@4}$	$1_{@2}$	$1_{@3}$	$0_{@1}$	$0_{@2}$	$2_{@4}$	$2_{@5}$	$1_{@3}$	$1_{@4}$	$0_{@2}$	$0_{@3}$	$2_{@5}$	$2_{@6}$
S(-2,4,10)	$0_{@0}$	∞	$0_{@1}$	∞	$2_{@3}$	∞	$1_{@2}$	∞	$0_{@1}$	∞	$3_{@5}$	∞	$2_{@4}$	∞	$1_{@3}$	∞

Table 2. Values and Stages for Some Additional Subtraction Games.

Horsefly

A **Horsefly**'s moves are 6 of the 8 possible moves of a knight in Chess (see Fig. 11) but there's another restriction: it can never exactly reverse the last move it made (even if other horseflies have moved meanwhile). The game shown in Fig. 11 is won by whoever lands the first horsefly at a winning post. Would you like to play first?

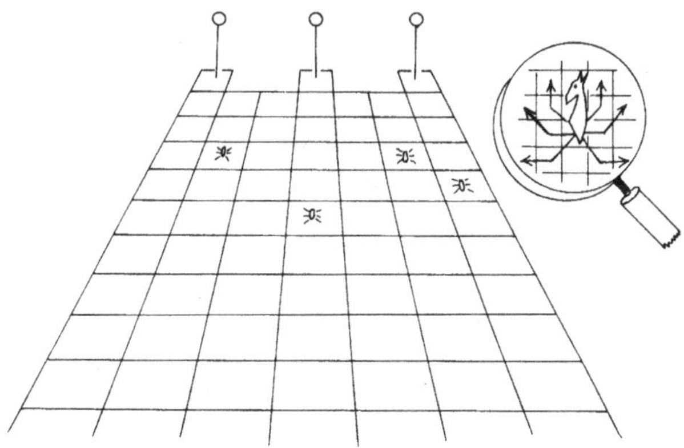

Figure 11. A Game of Horsefly.

Selective and Subselective Compounds of Impartial Games

In the same paper that deals with loopy impartial games, C. A. B. Smith discusses selective compounds in the impartial case. The answers are very easy. In the normal play case,

$$G \vee H \vee \ldots$$

is a $\mathcal{P}$-position only when all of

$$G, H, \ldots$$

are. The misère play outcome coincides with the normal outcome unless there is only one game that has not yet ended.

Smith has also discussed what we shall call **subselective compounds** in which the legal move is to move in any selection of components *except* the whole set. Such a compound is a $\mathcal{P}$-position only when the nim-values of all its components are equal. (If they're *not* all equal, you should find the lowest and reduce the rest to coincide with it.)

Entailing Moves

Top Entails is a new type of heap game, played with stacks of coins. You must either split a stack into two smaller ones or remove the top coin from a stack. In the latter case your opponent's next move must use the same stack. Don't leave a stack of 1 on the board, because your opponent can remove it and then demand that you follow with a move in the resulting empty stack!

The theory of sums of games when such **entailing moves** are allowed appears for the first time in *Winning Ways*. In such sums

$$A + B + C + \ldots$$

you may make a move in any component, unless your opponent has just made an *entailing move* somewhere, when you are required to follow him with a move in the same component. Often the best reply to one entailing move is another, and we might have a sustained rally of entailing moves in the same component.

The theory of such games reduces to Nim in a new way. The typical component A has a whole collection

$$a_0 < a_1 < a_2 < a_3 < \ldots$$

of nim-values, only the *least* of which, a_0, is **relevant**, unless the component has just been reached by an entailing move, when *all* are relevant. To compute new values:

> The complete set of nim-values for G
> consists of *all* numbers
> $g_0 < g_1 < g_2 < \ldots$
> *not* among the relevant values for options of G.
> Only g_0 is relevant
> unless the move to G is entailing.

And to find good moves:

> A move to
> $A + B + C + \ldots$
> that *entails* a reply in component A
> is good whenever there is *any* a_i for which
> $*a_i + *b_0 + *c_0 + \ldots = 0$
> while a *non*-entailing move to this position
> is good, as usual, just when
> $*a_0 + *b_0 + *c_0 + \ldots = 0.$

To see that these rules work, if the current position

$$A + B + C + \ldots = G,$$

of total value

$$*a_0 + *b_0 + *c_0 + \ldots = *g_0,$$

is not entailed, then, much as in the ordinary theory, the nim-values of its options include every number less than g_0, but not g_0 itself. If the move to A was entailing, the total values after possible moves are precisely those nimbers *not* of the form

$$*a_i + *b_0 + *c_0 + \ldots.$$

Here is a table of values for Top Entails:

0	⊙	10	$*1,4 \to$	20	$*2,5,8 \to$	30	$*2,7 \to$	40	$*1,4,8,9,12 \to$
1	☽	11	$*3.$	21	$*0,1,7.$	31	$*0,5.$	41	$*6,7,11.$
2	⊙	12	$*4 \to$	22	$*3,4,6,8 \to$	32	$*8 \to$	42	$*2,5,8,10,12 \to$
3	☽	13	$*0,2.$	23	$*2,5.$	33	$*1,6.$	43	$*9.$
4	$*1 \to$	14	$*3,5 \to$	24	$*7 \to$	34	$*5,9 \to$	44	$*3,6,7,8,11,13 \to$
5	$*0.$	15	$*4.$	25	$*3.$	35	$*0,7,8.$	45	$*0,2,4,10,12.$
6	$*2 \to$	16	$*2,6 \to$	26	$*1,4,6,8 \to$	36	$*1,6,10 \to$	46	$*5,7,8,9,13 \to$
7	$*1.$	17	$*5.$	27	$*7.$	37	$*4,5,9.$	47	$*3,6,10,11.$
8	$*3 \to$	18	$*1,3,4,7 \to$	28	$*3,5,8 \to$	38	$*7,8,11 \to$	48	$*4,8,12 \to$
9	$*0,2.$	19	$*6.$	29	$*4,6.$	39	$*10.$	49	$*2,7,9,10.$

Sunny and Loony Positions

In our tables

$$*n \to$$

denotes all nimbers from $*n$ onwards, and

$$*\bar{n}$$

all nimbers other than $*n$.

So

$*5 \rightarrow$	means	$*5, *6, *7, *8, *9, \ldots$
$*\bar{5}$	means	$0, *1, *2, *3, *4, *6, *7, *8, \ldots$
$*0, 1, 7.$	means	$0, *1, *7$
and $*3, 4, 6, 8 \rightarrow$	means	$*3, *4, *6, *8, *9, *10\ldots.$

But we'll often use

$$\odot \quad (\text{"sunny"})$$

instead of $*0 \rightarrow$ for the collection

$$0, *1, *2, *3, *4, \ldots$$

of *all* nimbers, and

$$☽ \quad (\text{"loony"})$$

for the empty collection, of *no* nimbers. What do these mean in practice?

When we defined Top Entails we warned you not to leave a stack of 1 because your opponent could remove it and demand that you follow with a move in the resulting empty stack. Your move would be *loony*, his *sunny*. In general:

> A **sunny move** in a component is one that wins for its maker, and
> a **loony move** is one that loses for him,
> *no matter what other components there are.*

In Top Entails both possible moves from a stack of 2 leave at least one stack of 1, so a move which entails your opponent to move from a stack of 2 is sunny. It's therefore loony to leave a stack of 3 (entailed or not) because your opponent can reply with this sunny move.

Remember that

$$\odot \text{ means } 0, *1, *2, *3, \ldots,$$

and that only the *earliest* value is relevant for *non*-entailing moves, so a stack of 2 reached by an ordinary move has the ordinary value 0.

Calculating with Entailed Values

Figure 12 shows how we found the values for Top Entails. The Grundy scale is set to compute the values for a 14-stack, S_{14}. For the non-entailing moves to

$$S_1 + S_{13}, \quad S_2 + S_{12}, \quad S_3 + S_{11}, \quad S_4 + S_{10}, \quad S_5 + S_9, \quad S_6 + S_8, \quad S_7 + S_7$$

only the earliest values are relevant:

$$☽ + 0, \quad 0 + *4, \quad ☽ + *3, \quad *1 + *1, \quad 0 + 0, \quad *2 + *3, \quad *1 + *1,$$

yielding

$$☽,\qquad *4,\qquad ☽,\qquad 0,\qquad 0,\qquad *1,\qquad 0,$$

while the entailed option S_{13} has two values:

$$0,\qquad *2.$$

The values for S_{14} are the remaining nimbers

$$*3, *5, *6, *7, *8, \ldots$$

and so we should write $3, 5 \rightarrow$ in the next place on the lower scale, and 3 on the upper one.

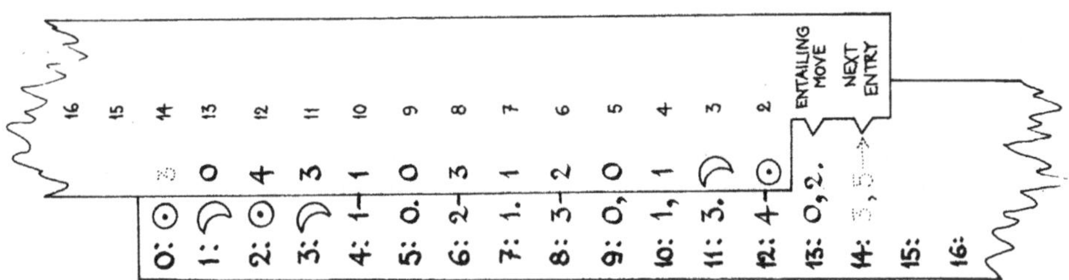

Figure 12. Grundy Scale Set to Calculate S_{14}'s Values.

The position

$$S_2 + S_6 + S_{14}$$

has value

$$0 + *2 + *3 = *1,$$

so there should be a good move reducing S_{14} to value $*2$. Our analysis shows that this must be the entailing move to S_{13}. If we make this, our opponent is forced to replace S_{13} by one of

$$S_1 + S_{12},\quad S_2 + S_{11},\quad S_3 + S_{10},\quad S_4 + S_9,\quad S_5 + S_8,\quad S_6 + S_7\quad \text{and } S_{12} \text{ (entailing)}$$

whose values

$$☽,\quad 0 + *3,\quad ☽,\quad *1 + 0,\quad 0 + *3,\quad *2 + *1,\quad *4, *5, *6, \ldots$$

do not include $*2$, so the values for the whole position will not include 0.

Of course, since ☽ represents the empty set, we have the obvious addition rules

$$☽ + *n = ☽ \qquad (n = 0, 1, 2, \ldots),$$
$$☽ + ☽ = ☽.$$

We don't know if Top Entails contains a loony stack of more than 3 coins. The first few single-stack $\mathcal{P}$-positions are

$$S_0, S_2, S_5, S_9, S_{13}, S_{21}, S_{31}, S_{35}, S_{45}, S_{57}.$$

Nim with Entailing Moves

It's easy to analyze any version of Nim in which we declare that some of the moves are to be entailing. As our first example we'll suppose that a move is entailing only when it decreases a heap by exactly 1. Even the move that replaces a heap of size 1 by the empty heap is entailing; since the other player cannot follow it, such a move is an outright win.

The values have an obvious pattern:

0	1	2	3	4	5	6	7	8	...
(◉ =) $*0 \to$	☽	$*1 \to$	☽	$*2 \to$	☽	$*3 \to$	☽	$*4 \to$	...

To see why this pattern continues, observe that a heap of size 9 is loony because we have non-entailing moves to heaps of sizes 0, 2, 4 and 6, values $0, *1, *2$ and $*3$, and an entailing move to a heap of size 8, values $*4 \to$, covering every possibility. And a heap of size 10 has values $*5 \to$ because the only non-loony moves are non-entailing, to heaps of sizes $0, 2, 4, 6$ or 8, values $0, *1, *2, *3$ or $*4$.

When the move which empties a heap of size 1 is declared *non*-entailing, it's no longer loony to leave a 1-heap, and the rest of the pattern shifts, making the *even* heaps loony:

0	1	2	3	4	5	6	7	8	9	10	11	12	...
$*0 \to$	$*1 \to$	☽	$*2 \to$	☽	$*3 \to$	☽	$*4 \to$	☽	$*5 \to$	☽	$*6 \to$	☽	... .

We might instead play the game $N(a, b, c, ...)$ which is Nim in which moves that reduce the size of a heap by a or b or c or ... are declared entailing (even when the next player is unable to move in the reduced heap). The theory is easily deduced from that of the corresponding Subtraction Game $S(a, b, c, ...)$ (see Chapter 4).

Here are the value sequences for $S(2, 5, 7)$ and $N(2, 5, 7)$:

$S(2,5,7)$	$\dot{0}$.	0	1	1	0	2	1	3	2	2	0	3	1	0	0	1	1	2	2	3	3	$\dot{2}$
n	0	1	2	3	4	5	6	7	8	9	10	11	12	13	14	15	16	17	18	19	20	21
$N(2,5,7)$	$*0 \to$	$*1 \to$	☽	☽	$*2 \to$	☽	☽	☽	☽	☽	$*3 \to$	☽	☽	$*4 \to$	$*5 \to$	☽	☽	☽	☽	☽	☽	☽
n	22	23	24	25	26	27	28	29	30	31	32	33	34	35	36	37	38	39	40	41	...	
$N(2,5,7)$	$*6 \to$	$*7 \to$	☽	☽	$*8 \to$	☽	☽	☽	☽	☽	$*9 \to$	☽	☽	$*10 \to$	$*11 \to$	☽	☽	☽	☽	☽	....	

You will see that every non-zero value of $S(2, 5, 7)$ becomes loony in $N(2, 5, 7)$ and that the remaining values for $N(2, 5, 7)$ are just

$$*0 \to *1 \to \quad *2 \to \quad *3 \to \quad *4 \to \ldots$$

in order.

For a heap H in such a game, if the non-loony values of smaller heaps are

$$*0 \to, *1 \to, *2 \to, \ldots, *(n-1) \to,$$

then

$$*0, *1, *2, \ldots, *(n-1)$$

certainly appear among the options of H. If none of the corresponding moves is entailing, the values of H are therefore $*n \rightarrow$, but if any of these moves *is* entailing, then *all* nimbers appear among the option values, so H will be loony.

For example, the non-loony option values for a heap of size 27 in $N(2,5,7)$ are

$$*0, *1, *2, *3, *4, *5, *6 \rightarrow, *7, *8$$

(since the move to a heap of size 22 is entailing), so this heap is loony.

In general

> in a Nim game with any collection of moves declared entailing, the values are $*0 \rightarrow, *1 \rightarrow, *2 \rightarrow, *3 \rightarrow, \ldots$ in order, interspersed with blocks of loony positions.

If we modify $N(2,5,7)$ by *dis*entailing the moves to 0-heaps we obtain

n	0	1	2	3	4	5	6	7	8	9	10	11	12	...
	⊙	$*1 \rightarrow$	$*2 \rightarrow$	☽	☽	$*3 \rightarrow$	☽	☽	☽	☽	☽	$*4 \rightarrow$	☽	...

and if we also disentail moves to 1-heaps:

n	0	1	2	3	4	5	6	7	8	9	10	11	12	...
	⊙	$*1 \rightarrow$	$*2 \rightarrow$	$*3 \rightarrow$	☽	☽	$*4 \rightarrow$	☽	☽	☽	☽	☽	$*5 \rightarrow$	... .

After their initial terms these patterns are obtained by displacing that for $N(2,5,7)$ (and increasing the nimbers by 1 or 2).

Goldbach's Nim

The same kind of argument works no matter what exotic conditions are required for a move to be entailing. **Goldbach** declares that a move is entailing just when it both takes a prime and leaves a prime. If we don't count 1 as a prime, the first few non-loony values are, for $n =$

0	1	2	3	11	12	17	23	27	29	35	37	38	...
⊙	$*1 \rightarrow$	$*2 \rightarrow$	$*3 \rightarrow$	$*4 \rightarrow$	$*5 \rightarrow$	$*6 \rightarrow$	$*7 \rightarrow$	$*8 \rightarrow$	$*9 \rightarrow$	$*10 \rightarrow$	$*11 \rightarrow$	$*12 \rightarrow$	...

But Goldbach considered 1 to be prime and so for his form of the game we find that the only non-loonies are

0	1	5	9	11	15	17	21	23	27	29	33	35	37	...
⊙	$*1 \rightarrow$	$*2 \rightarrow$	$*3 \rightarrow$	$*4 \rightarrow$	$*5 \rightarrow$	$*6 \rightarrow$	$*7 \rightarrow$	$*8 \rightarrow$	$*9 \rightarrow$	$*10 \rightarrow$	$*11 \rightarrow$	$*12 \rightarrow$	$*13 \rightarrow$	...

For this game we can prove that an odd number $n > 7$ is loony just if $n-2$ is prime and $n-4$ is not. We are sure that Goldbach believes that all even numbers are loony: an even number n is non-loony only if for *every* expression

$$n = p + q$$

of n as the sum of 2 primes, each of p and q is either 3 or the larger member of a prime pair. A near miss was 122; can you find its unique entailing move to a non-loony position?

Wyt Queens with Trains

If you don't remember the game of Wyt Queens, look at Fig. 5 on p. 60 in Chapter 3. The Queens moved either orthogonally or diagonally towards the corner of the board. Let's see what happens when some of the moves are declared entailing. Table 3 shows the values when we declare that just the orthogonal moves are entailing. The zeros are in the same places as in the ordinary game and each begins a diagonal trail, $0, 1, 2, 3, \ldots$; apart from these the values are loony.

	0	1	2	3	4	5	6	7	8	9	10	11
0	0→											
1		1 →	0.									
2		0.	2 →	1.					☽			
3			1.	3 →	2.	0.						
4				2.	4 →	3.	1.	0.				
5				0.	3.	5 →	4.	2.	1.			
6		☽			1.	4.	6 →	5.	3.	2.	0.	
7					0.	2.	5.	7 →	6.	4.	3.	1.

Table 3. Entailing Orthogonally and Trailing Diagonally.

If instead we declare that just the diagonal moves are entailing we get a much more interesting game, **Off-Wyth-Its-Tail!**, whose values appear in Table 5. This time the non-loony values except zero form corridor-like trails in the two orthogonal directions. The zero values would begin the corridors if they were displaced one diagonal place.

There are lots of changes you can make to the entailing rules in this game which don't affect the values. So long as you

don't entail orthogonal moves *along* the boundary, and
do entail diagonal moves *onto* the boundary,

the other moves may be entailing or not, and the values will stay the same.

nim-value	k:	0	1	2	3	4	5	6	7
difference d:	0	(0,0)							
	1	(1,2)	(0,1)						
	2	(3,5)	(2,4)	(0,2)					
	3	(4,7)	(3,6)	(1,4)	(0,3)				
	4	(6,10)	(5,9)	(3,7)	(1,5)	(0,4)			
	5	(8,13)	(7,12)	(5,10)	(2,7)	(1,6)	(0,5)		
	6	(9,15)	(8,14)	(6,12)	(4,10)	(2,8)	(1,7)	(0,6)	
	7	(11,18)	(10,17)	(8,15)	(6,13)	(3,10)	(2,9)	(1,8)	(0,7)
	8	(12,20)	(11,19)	(9,17)	(8,16)	(5,13)	(3,11)	(2,10)	(1,9)

Table 4. Coordinates of Values in Table 5.

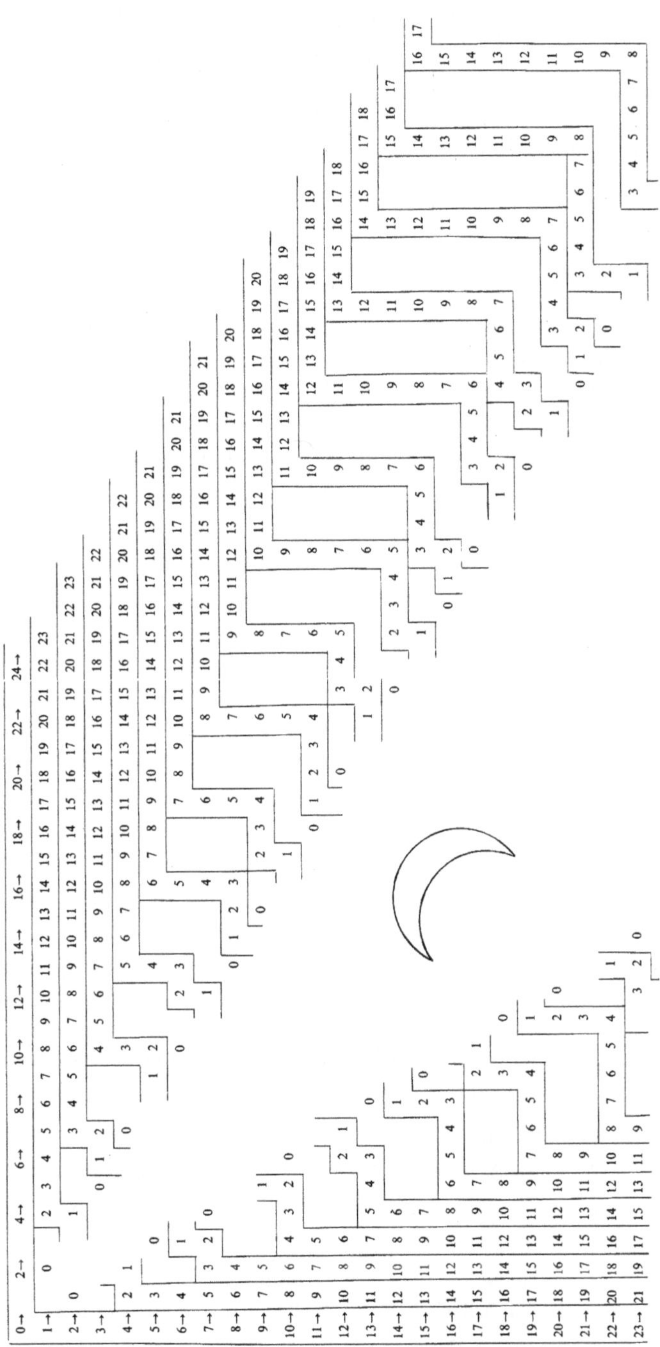

Table 5. Entailing Diagonally and Trailing Orthognally.

The Difference Rule that we gave in the Extras to Chapter 3 for the $\mathcal{P}$-positions in Wythoff's game extends to a rule for all non-loony values in Off-Wyth-Its-Tail! Any desired k is the nim-value of infinitely many positions (x, y), one for each difference $d \geq k$, as in Table 4. The pair with difference d in any column of this table is found as $(x, x + d)$ where x is the smallest number that doesn't appear in the higher pairs of that column. Write to us if you can prove all our statements about this game in a few lines—our proofs are longer.

Adding Tails to Prim and Dim

We met Prim and Dim in Chapter 4. In Prim you may remove m beans from a heap of n provided m and n have no common factor bigger than 1 ("provided m and n are *coprime*"). We shall now add the condition that when one player reduces a heap by 1, his opponent must follow in the same heap. In other words, *reduction by* 1 *is an entailing move.*

There are several cases: in the first version, reducing 1 to 0 is a *legal* move. If this move is declared *entailing* we obtain rather trivial nim-values:

$n =$	0	1	2	3	4	5	6	7	8	9	10	...
	⊙	☽	⊙	☽	⊙	☽	⊙	☽	⊙	☽	⊙	....

If 1 to 0 is *legal* but not entailing we find:

$n =$	0	1	2	3	4	5	6	7	8	9	10	11	12	13	14
	$*0$	$*1\to$	$*0$	$*2\to$	$*0$	$*3\to$	$*0,2$	$*4\to$	$*0$	$*2,5\to$	$*0,3$	$*5\to$	$*0,2$	$*6\to$	$*0,4$

$n =$	15	16	17	18	19	20	21	22	23	24	25	...
	$*2,3,7\to$	$*0$	$*7\to$	$*0,2$	$*8\to$	$*0,3$	$*2,4,9\to$	$*0,5$	$*9\to$	$*0,2$	$*3,10\to$	...

while if 1 to 0 is *illegal* and the move 2 to 1 is entailing:

$n =$	0	1	2	3	4	5	6	7	8	9	10	11	12	13	14	15	16	17
	–	⊙	☽	$*1\to$	☽	$*2\to$	$*1$	$*3\to$	☽	$*1,4\to$	$*2$	$*4\to$	$*1$	$*5\to$	$*3$	$*1,2,6\to$	☽	$*6\to$

$n =$	18	19	20	21	22	23	24	25	26	27	28	29	30	31	32	33	...
	$*1$	$*7\to$	$*2$	$*1,8\to$	$*4$	$*8\to$	$*1$	$*2,9\to$	$*5$	$*1,9\to$	$*3$	$*9\to$	$*1,2$	$*10\to$	☽	$*1,4,11\to$	...

In **Dim**, the move is to take a divisor off any heap, and now we shall add the condition that reduction by 1 is an entailing move. The values are:

$n =$	0	1	2	3	4	5	6	7	8	9	10	11	12	13	14
	⊙	☽	$*1\to$	☽	$*2\to$	$*1$	$*3\to$	$*1,2$	$*4\to$	$*1,2$	$*3,5\to$	$*1,2,4$	$*5\to$	$*1,2,3,4$	$*6\to$

$n =$	15	16	17	18	19	20	21	22	23	24
	$*1,2,4$	$*3,7\to$	$*1,2,4,5,6$	$*7\to$	$*1,2,3,4,5,6$	$*8\to$	$*1,2,3,4,5$	$*6,7,9\to$	$*1,2,3,4,5,8$	$*9\to$

$n =$	25	26	27	28	29	30	31	32	...
	$*1,2,3,4,5,6,7$	$*8,10\to$	$*1,2,3,4,5,6$	$*7,10\to$	$*1,2,3,4,5,6,8,9$	$*10\to$	$*1,2,3,4,5,6,7,8,9$	$*11\to$	

If the move n to 0 is made illegal and that from 2 to 1 is not entailing, $*0$ is adjoined to the values of all *odd* heaps.

Complimenting Moves

In some games there are special **complimenting moves**, after which the same player has an extra **bonus move**. This happens in Dots-and-Boxes (Chapter 16) when a box is completed and in many children's board games when a double-six is thrown (to him that hath shall be given). The bonus move is quite free and can even be another complimenting move. One player may be twiddling his thumbs for quite a while while his opponent takes a whole string of consecutive moves. We can think of these thumb-twiddles as entailed pass moves whose only function is to make the turn alternate.

Let's cook up an example. **All Square** is a heap game in which the move is to split any heap into two smaller ones, and the complimenting moves are just those for which the heap sizes of these smaller ones are both perfect squares. When you've made a complimenting move you *must* move again—if you can't do so you lose—a *back-handed* compliment! Here are the values for All Square:

$n =$ 1	2	3	4	5	6	7	8	9	10	11	12	13	14	14	16	17
$\boxed{0}$	0	$*1$	$\boxed{*2}$	$*2$	$*1$	0	☽	$\boxed{*1}$	☽	0	$*1$	$*3$	$*2$	$*1$	$\boxed{0}$	0
$n =$ 18	19	20	21	22	23	24	25	26	27	28	29	30	31	32	33	34
☽	$*4$	$*2$	$*1$	$*3$	$*2$	$*1$	$\boxed{☽}$	☽	$*3$	$*4$	☽	$*1$	0	☽	$*4$	☽
$n =$ 35	36	37	38	39	40	41	42	43	44	45	46	47	48	49	50	51
$*5$	$\boxed{*3}$	$*3$	$*2$	$*1$	☽	☽	$*3$	$*4$	$*1$	☽	$*3$	$*2$	$*1$	$\boxed{*6}$	☽	$*3$
$n =$ 52	53	54	55	56	57	58	59	60	61	62	63	64	65	66	67	68
☽	$*4$	$*6$	$*3$	$*2$	$*4$	☽	$*2$	$*3$	☽	$*2$	$*8$	$\boxed{*6}$	☽	$*4$	$*1$	☽
$n =$ 69	70	71	72	73	74	75	76	77	78	79	80	81	...			
$*3$	$*5$	$*4$	☽	☽	☽	0	$*1$	$*7$	$*3$	$*5$	☽	$\boxed{*4}$	...			

To help you to recognize the complimenting moves, we've put square boxes round the values for square numbers. We found the values by using the following principle:

> A complimenting move
> to a position of value $*n$
> is really a move to
> $*\bar{n} = *0, *1, ..., *(n-1), *(n+1) \rightarrow,$
> i.e., every nimber *other* than $*n$.

THE COMPLEMENTING EFFECT OF COMPLIMENTING MOVES

For after such a move, the opponent's move is entailed—he must twiddle his thumbs and then leave a position of value $*n$. Our rules for entailing moves now show that what we left him had all values $*\bar{n}$.

In Figure 13 we see a Grundy scale for All Square, set for a heap of 13; and below it we have written the values of the various options. Since the move to 9 + 4 is complimenting, it

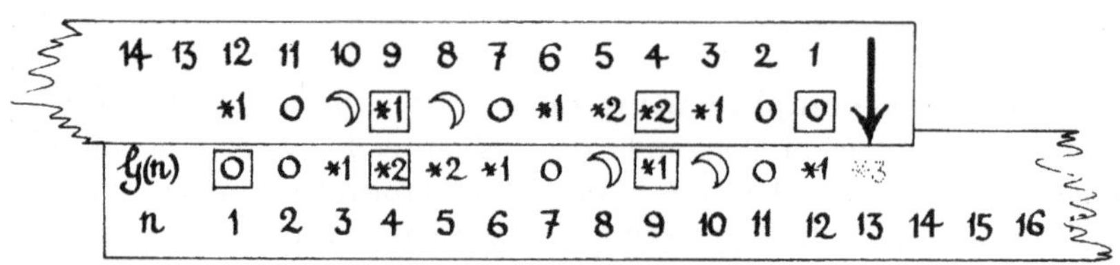

Figure 13. What's a Heap of 13 Worth?

alone accounts for all nimbers other than $*3$, and since $*3$ does not appear elsewhere it is the answer.

Why is 58 a loony? We can see this using only the values

$$9 : \boxed{*1}, \qquad 49 : \boxed{*6}, \qquad 23 : *2, \qquad 35 : *5.$$

The complimenting move to $9+49$ accounts for all nimbers other than $*7$, which is catered for by the ordinary move to $23+35$. What does this mean in practice? If your opponent leaves you a position

$$58 + A + B + \ldots,$$

move to

$$9 + 49 + A + B + \ldots$$

unless the value of $A + B + \ldots$ is $*7$. The total position then has non-zero value and you can use your bonus move to correct it. If the value of $A + B + \ldots$ *is* $*7$, move instead to

$$23 + 35 + A + B + \ldots .$$

This is a kind of strategy stealing. We find out who has the winning strategy in

$$*7 + A + B + \ldots$$

and make arrangements to appropriate it for ourselves.

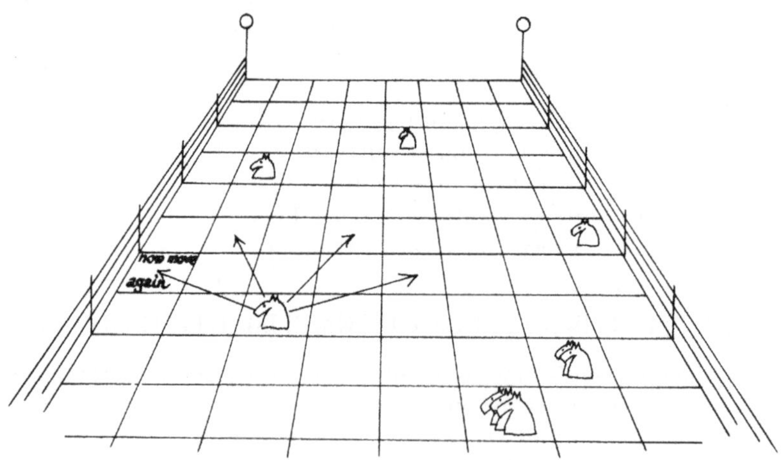

Figure 14. Horse Moves for On-the-Rails.

On-the-Rails

On-the-Rails is a racecourse game with complimenting moves. The horses move as shown in Fig. 14, there can be several on a square; and the player who first moves a horse across the finishing line is the winner.

But this time there's a new twist—if you put a horse against the rails you take an extra move (which need not involve the same horse). In other words a move onto a square in either the leftmost or rightmost column is a complimenting move.

The values are easy to analyze and ultimately have period 7 as shown in Table 6. In each square we show the immediate value(s) of a move onto that square. If the horse is left on the rails after the corresponding bonus move, its value is found by dropping the bar.

⊙	☽	☽	☽	☽	☽	☽	⊙
⊙	☽	☽	☽	☽	☽	☽	⊙
$*\bar{0}$	☽	☽	$*0$	$*0$	☽	☽	$*\bar{0}$
$*\bar{0}$	☽	☽	$*0$	$*0$	☽	☽	$*\bar{0}$
$*\bar{0}$	☽	☽	$*1$	$*1$	☽	☽	$*\bar{0}$
$*\bar{0}$	$*0$	☽	$*1$	$*1$	☽	$*0$	$*\bar{0}$
$*\bar{0}$	$*0$	$*0$	$*2$	$*2$	$*0$	$*0$	$*\bar{0}$
$*\bar{1}$	$*0$	☽	$*2$	$*2$	☽	$*0$	$*\bar{1}$
$*\bar{1}$	☽	$*1$	$*1$	$*1$	$*1$	☽	$*\bar{1}$
$*\bar{2}$	☽	☽	$*0$	$*0$	☽	☽	$*\bar{2}$
$*\bar{0}$	☽	$*2$	$*0$	$*0$	$*2$	☽	$*\bar{0}$
$*\bar{0}$	$*2$	☽	$*1$	$*1$	☽	$*2$	$*\bar{0}$
$*\bar{0}$	$*0$	☽	$*1$	$*1$	☽	$*0$	$*\bar{0}$
$*\bar{0}$	$*0$	$*0$	$*2$	$*2$	$*0$	$*0$	$*\bar{0}$
$*\bar{1}$	$*0$	☽	$*2$	$*2$	☽	$*0$	$*\bar{1}$
$*\bar{1}$	☽	$*1$	$*1$	$*1$	$*1$	☽	$*\bar{1}$
$*\bar{2}$	☽	☽	$*0$	$*0$	☽	☽	$*\bar{2}$
$*\bar{0}$	☽	$*2$	$*0$	$*0$	$*2$	☽	$*\bar{0}$
$*\bar{0}$	$*2$	☽	$*1$	$*1$	☽	$*2$	$*\bar{0}$

Table 6. Values for On-the-Rails.

Extras

De Bono's *L*-Games

N. E. Goller has found a simple and elegant strategy by which either player can guarantee to draw (at least) from the initial position of Fig. 5 (and from many others). Place your *L*-piece

either so that it occupies *three* of the central four squares (Fig. 15(a))
or so that it occupies *two* of those squares, *and* no neutral piece
occupies any other squares marked X in Fig. 15(b).

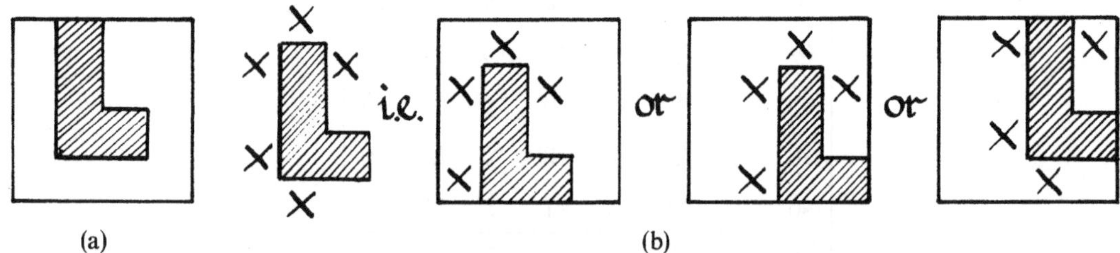

Figure 15. Nick Goller's Drawing Strategy for the *L*-Game.

Figure 16 shows a longest sensible finishing game, assuming that both players are playing well.

Proving the Outcome Rules for Loopy Positions

For the **closed** ($\mathcal{N}$ or $\mathcal{P}$) positions, this is easy. The only closed case not covered by the Nim theory is the assertion that

$$\infty_{abc\ldots} + *n \text{ is } \mathcal{N} \text{ if } n \text{ is one of } a, b, c, \ldots.$$

However, from this position the next player can move to

$$*n + *n = 0.$$

The other positions are of two types:

(i) $\infty_{abc\ldots} + *n$ with n *not* one of $a, b, c, \ldots,$

(ii) $\infty_{abc\ldots} + \infty_{\alpha\beta\gamma\ldots} + \ldots.$

Because we don't yet know if these are open or closed, we'll call them **ajar** for the moment. We need only show that such a position has an option which is still ajar.

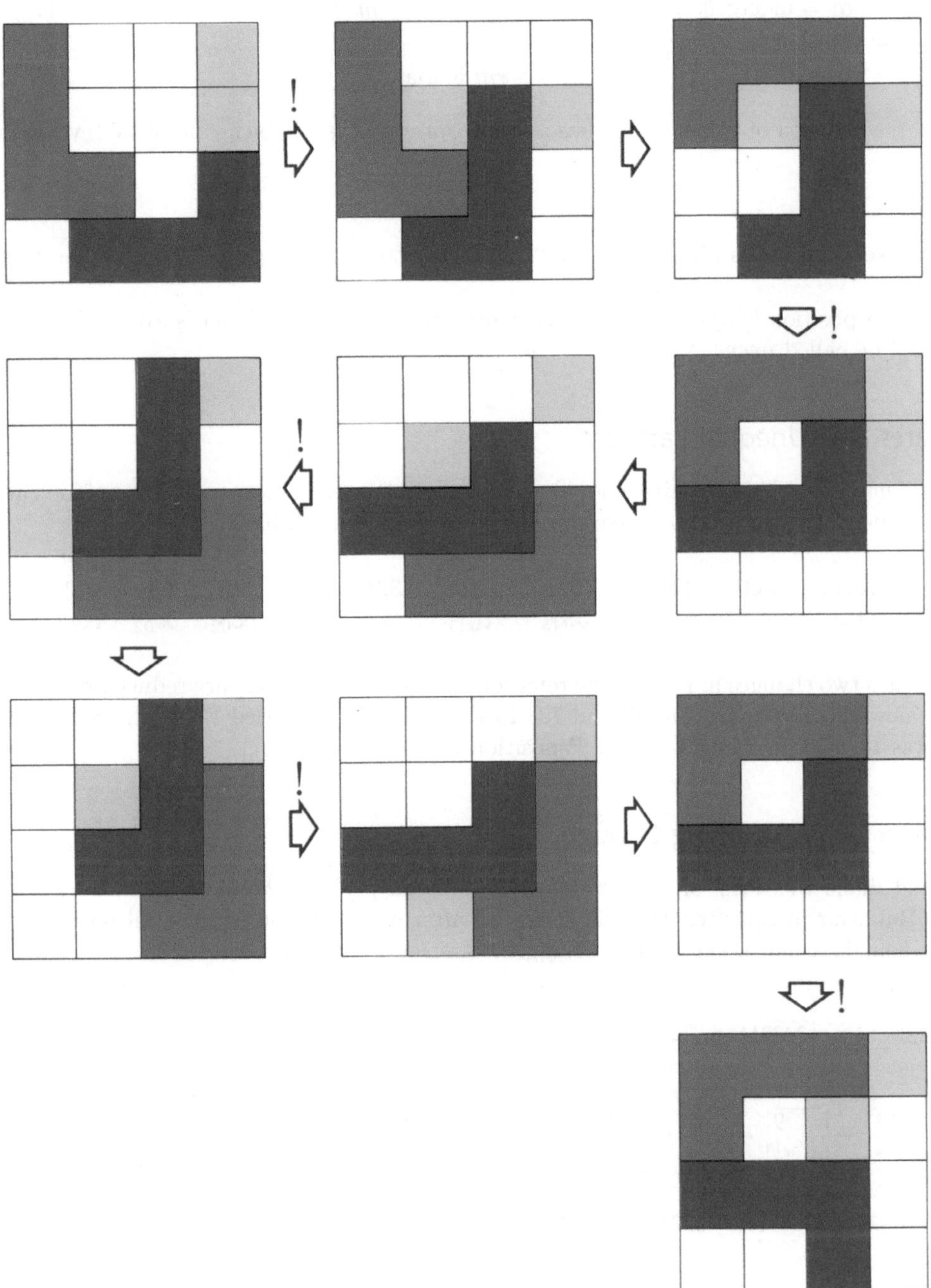

Figure 16. A Longest Sensible Finishing Game.

For (i), let $m = \text{mex}(a, b, c, ...)$ and observe that $n \geq m$ because n is not one of $a, b, c, ...$ If $n > m$, the position

$$\infty_{abc...} + *m \text{ is ajar.}$$

If $n = m$, the position of value $\infty_{abc...}$ has a loopy option $\infty_{\alpha\beta\gamma...}$ which cannot be reversed to $*n$, so

$$\infty_{\alpha\beta\gamma...} + *n \text{ is ajar.}$$

We can keep positions of type (ii) ajar by moving to a loopy option of one of the loopy components.

So when a position is ajar, either player can keep it ajar. A door that's always kept ajar may as well be called *open.*

Fair Shares and Unequal Partners

If, in Fair Shares and Varied Pairs, you allow yourselves the extra moves which combine three or more heaps, then you need to make the following changes to Table 1:

421	32111	3321	42111	4221	441	4321	532	541	631	721
$*4_{@6}$	$*4_{@11}$	$*1_{@10}$	$*4_{@12}$	$*5_{@15}$	$*1_{@11}$	∞_{02}	∞_{02}	∞_{02}	∞_{02}	∞_{02} .

There are also two changes in Fig. 4: the remotenesses of 4221 and 4321 are reduced to 3, since there are now the good moves to 72 and 73. In Fair Shares and Varied Pairs, the remoteness of 4221 was 14, so this changes from a $\mathcal{P}$-position into an $\mathcal{N}$-position.

Were Your Ways Winning Enough?

You should make the date, of course, since the almond position 22221 is labelled $*2_{@12}$ in Table 1. But your prospective date might cry off after watching you swallow all but 2 of the toffees at your first move, and then dragging the game out to 25 moves in all:

22221 + $*n$ → [22221 + $*2$] → 22221 + $*1$ / 3222 + $*2$ → [3222 + $*1$] → 32211 + $*1$ / 522 + $*1$ → [2211111 + $*1$]

→ 21111111 + $*1$ ← [3111111 + $*1$] ← 411111 + $*1$ ← [51111 + $*1$] ← 6111 + $*1$ ← [711 + $*1$] ← 81 + $*1$ ← [9 + $*1$] ← 333 + $*1$

→ [333] → 33111 → [4311] → 32211 → [5211] → 531 → [54] → 522 → [72] → 9 / 711 / 21111111 → [111111111]

(This is your only way to force a win so quickly.) There are times when your *Winning Ways* should not be used too blatantly!

Did You Move First in Horsefly?

You should have chosen *not* to move first, since the position of Fig. 11 is a $\mathcal{P}$-position, no matter which way those horseflies last moved. In Fig. 17 the usual value of a square is printed at its centre. However, if you've just arrived in the third or fifth column by a knight's move from the other, then you should use the small print value in the corner nearest to where you came from.

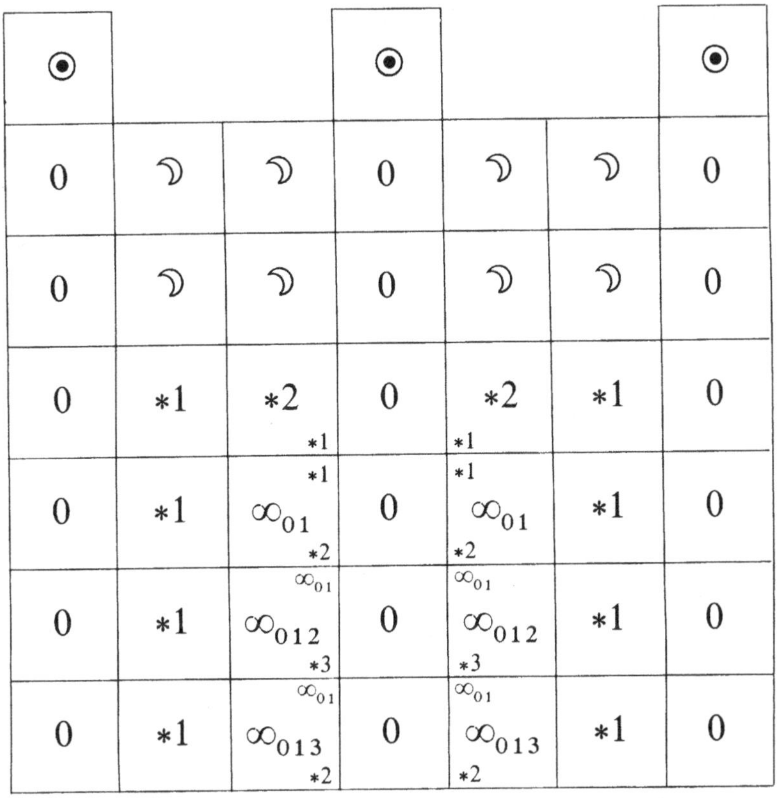

⊙			⊙			⊙
0	☽	☽	0	☽	☽	0
0	☽	☽	0	☽	☽	0
0	$*1$	$*2$ $*1$	0	$*2$ $*1$	$*1$	0
0	$*1$	$*1$ ∞_{01} $*2$	0	$*1$ ∞_{01} $*2$	$*1$	0
0	$*1$	∞_{01} ∞_{012} $*3$	0	∞_{01} ∞_{012} $*3$	$*1$	0
0	$*1$	∞_{01} ∞_{013} $*2$	0	∞_{01} ∞_{013} $*2$	$*1$	0

Figure 17. Values for Horsefly.

We've used

⊙ for the three winning posts,
☽ for squares from which there is a move to ⊙,

and to follow the later analysis you should fill in the remaining values in the order

$$0, *1, *2, *3, \infty_{01}, \infty_{012}, \infty_{013},$$

proceeding row by row for each value. The last two rows repeat indefinitely.

References and Further Reading

Edward de Bono, *The Five-day Course in Thinking—Introducing the L-game*, Pelican, London 1969.

Edward de Bono, *The Use of Lateral Thinking*, Pelican, London 1967; Basic Books, N.Y., 1968.

A. S. Fraenkel and U. Tassa, Strategy for a class of games with dynamic ties, *Comput. Math. Appl.*, **1**(1975) 237–254; MR **54**#2220.

V. W. Gijlswijk, G. A. P. Kindervater, G. J. van Tubergen and J. J. 0. 0. Wiegerinck, Computer analysis of E. de Bono's L-game, Report #76–18, Dept. of Maths., Univ. of Amsterdam, Nov. 1976.

Cedric A. B. Smith, Graphs and composite games, *J. Combin. Theory*, **1**(1966) 51–81; MR **33** #2572.

-13-

Survival in the Lost World

"The game," he said, "is never lost till won."
George Crabbe, *Tales of the Hall*, xv, "Gretna Green", 334.

The genius of you Americans is that you never make clear-cut stupid moves, only complicated stupid moves which make us wonder at the possibility that there may be something to them we are missing.
Gamel Abdel Nasser

'Tis better to have fought and lost
Than never to have fought at all.
Arthur Hugh Clough, *Pescheria*

Exchewe allewey Caylys, carding and haserdy.
Advice to Apprentices: c. 1450.

We've spent a lot of time teaching you how to win games by being the last to move. But suppose you are baby-sitting little Jimmy and want, at least occasionally, to make sure you *lose*? This means that instead of playing the normal play rule in which whoever can't move is the *loser*, you've switched to the **misère play** rule when he's the *winner*. Will this make much difference? Not *always*...

Misère Nim

In **Misère Nim** you have a number of heaps and your move must reduce the size of any one heap, but whoever eliminates the last heap is now the *loser*. This will only affect play in the **fickle** positions when all the non-empty heaps are **singletons** (size 1) and the game gets rather boring and mechanical. You and Jimmy will take the singletons alternately and you would win in normal play by always presenting him with an even number of them. Obviously, to play in misère fashion, you should instead arrange to present Jimmy with an *odd* number of singletons. All other positions (some heaps ≥ 2) are **firm** and behave alike in both normal and misère play.

Whoever first reduces the game to the fickle state does so by reducing the last heap of size 2 or more. He can therefore win in either Normal or Misère Nim by choosing whether to reduce this heap to a singleton or remove it altogether.

> Play Misère Nim exactly as you would play Normal Nim unless your move would leave an even number of singleton heaps and no other heap. Then leave an odd number instead.

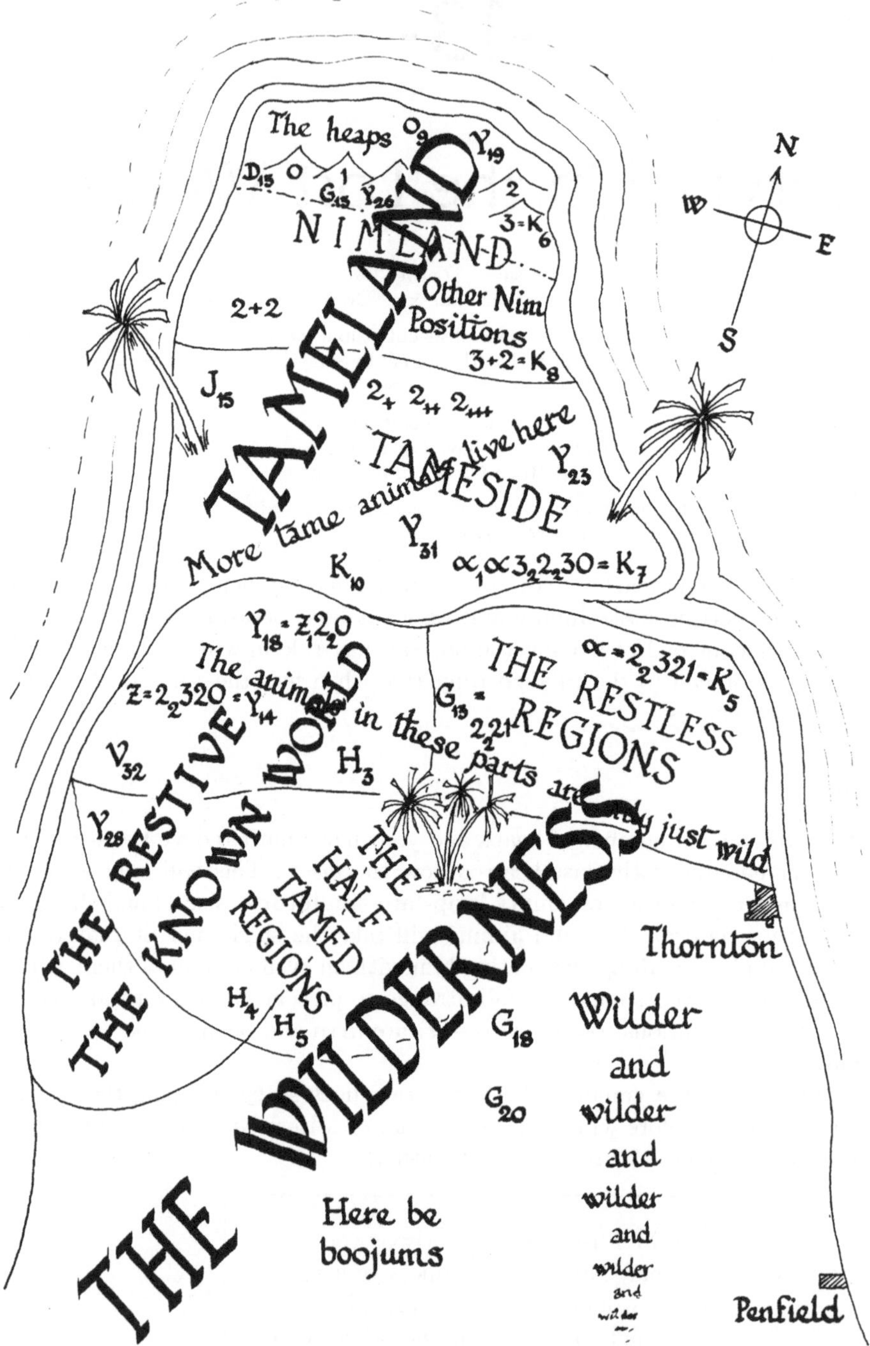

Figure 1. Game Reserves in the Lost World.

Reversible Moves

We can still throw out all reversible moves in the misère theory, provided we take special care when we throw *all* moves out. Figure 2, illustrating reversibility for misère play, is rather like Fig. 3 in Chapter 3, but *all* branches are now available to *either* player. The game G has all the options A, B, C of the simpler game H, and additional options D, E from each of which there is a legal move to H. We say that D and E are **reversible moves** from G.

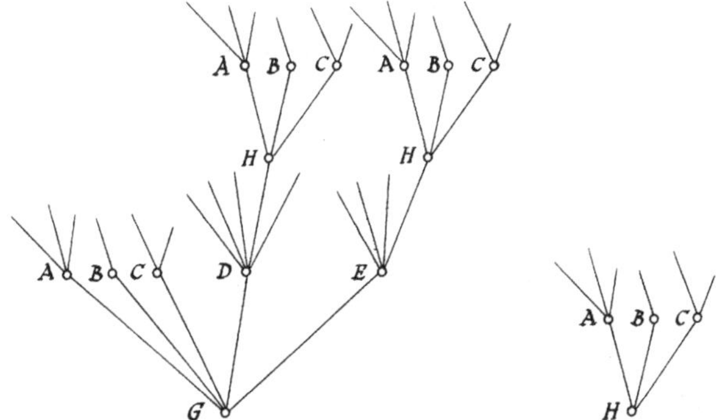

Figure 2. Reversible Moves in Misère Play.

We assert that the outcome of a sum of games will never be affected if we replace a component G by the simpler game H, provided H has at least one option. For, whoever has the winning strategy in

$$H + X + Y + Z + ...$$

can play

$$G + X + Y + Z + ...$$

with essentially the same strategy, not himself making use of the new moves from G to D or E. If his opponent uses them he can expect to reverse their effect by moving back from

$$(D \text{ or } E) + X + Y + Z + ...$$

to

$$H + X + Y + Z + ...$$

> If G is obtainable from H
> by adding reversible moves,
> then G is equivalent to H,
> *provided* that
> if H has no option,
> then G and H have the same
> outcome.

PRUNING REVERSIBLE MOVES

Since then we can substitute H for G in any sum of games we shall simply write $G = H$. For example,

$$\{0, 1, 2, 5, 6, 9\} = \{0, 1, 2\} = 3,$$

since 5, 6 and 9 have 3 as an option and 3 is not 0.

The Endgame Proviso

But this pruning process requires care when H has no legal option, i.e. is the **Endgame**, 0 (see Fig. 3). The trouble is that the argument might not even get off the ground.

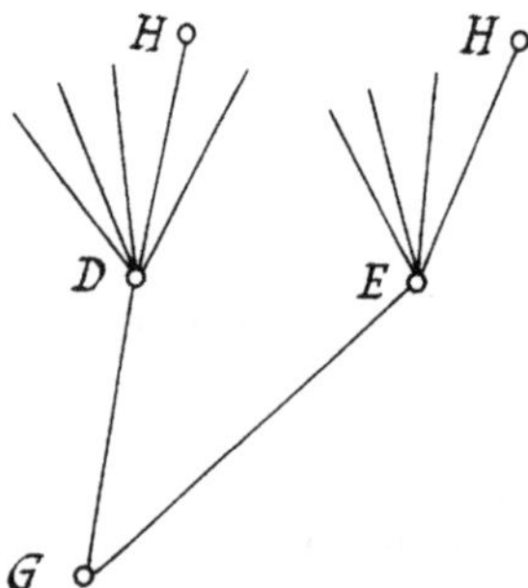

Figure 3. The Endgame Proviso

For if $H, X, Y, Z, \ldots$ are each the Endgame, you have already won

$$H + X + Y + Z + \ldots (= 0)$$

by definition, but this doesn't help you to win

$$G + X + Y + Z + \ldots (= G).$$

So in this case we are compelled to lay down the **proviso** that G has the *same outcome* as 0, when H *is* 0.

The Awful Truth

Since all impartial games reduce to Nim in normal play, and since misère play Nim is only a trivial modification of normal Nim, it has often been thought that misère impartial games must be almost as easy. Perhaps you play any sum of impartial games in misère play just as in normal play until very near the end, when ... ?

Unfortunately not!

For example, in the normal play version of Grundy's game (Chapter 4) the single heap $\mathcal{P}$-positions with less than 50 beans are

$$\mathrm{G}_1, \mathrm{G}_2, \mathrm{G}_4, \mathrm{G}_7, \mathrm{G}_{10}, \mathrm{G}_{20}, \mathrm{G}_{23}, \mathrm{G}_{26}.$$

None of these is $\mathcal{P}$ in misère play, for which the first few single heap $\mathcal{P}$-positions are

$$\mathrm{G}_3, \mathrm{G}_6, \mathrm{G}_9, \mathrm{G}_{12}, \ldots, \mathrm{G}_{42}, \mathrm{G}_{45},$$

exactly the first fifteen multiples of 3. It seems a coincidence that G_{50} is a $\mathcal{P}$-position in both kinds of play. Many authors have made mistakes in this and similar games by riding roughshod over the subtleties of misère play.

In ONAG it is proved that the *only* way a game can be simplified in misère play is by eliminating reversible moves, observing the Endgame Proviso when appropriate. If two games have no reversible moves and *look* different, they really *are* different, because there will always be some other game whose addition yields distinct outcomes.

The complications that arise are illustrated in Fig. 4 which shows the tree structure of the game of Kayles with 7 pins, simplified as much as possible.

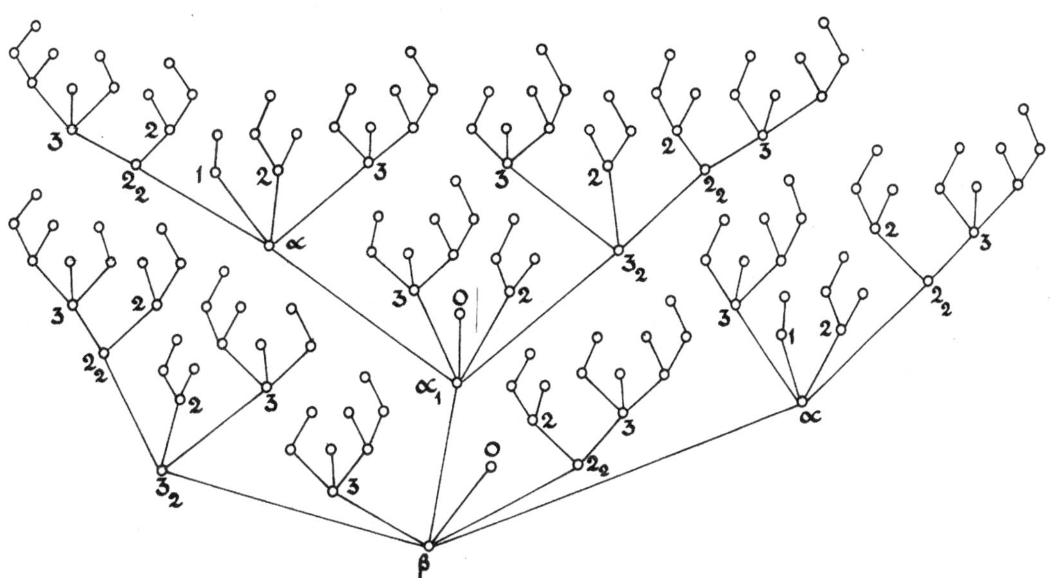

Figure 4. The Kayles Position K_7. Structure $\beta = \alpha_1\alpha 3_2 2_2 30$, Where $\alpha = 2_2 321$.

Grundy and Smith showed that there was 1 game born on day $0, 2$ by day 1, 3 by day $2, 5$ by day $3, 22$ by day 4 and 4171780 by day 5. So there are at most

$$2^{4171780}$$

games by day 6. Removal of reversible moves reduces this huge number, but by such a small fraction that the first 625140 of its 1255831 decimal digits are not affected! The exact number is given on p. 140 of ONAG.

We salvage what we can with our notion of **tame game**, but that's not too much. If you think the rest of the chapter looks rather complicated, that's because it is!

What's Left of the Old Rules?

Here, and for the rest of the chapter, we omit the stars from the notation for Nim-heaps, so that numbers will usually mean *nimbers.*

In normal play the mex rule could be used to reduce *every* position to a Nim-heap. Although we've just seen that this no longer happens, the rule still has some force:

> If $m = \text{mex}\,(a, b, c, \ldots)$ then
> $\{a, b, c, \ldots\} = m,$
> *provided* that at least one of $a, b, c, \ldots$ is 0 or 1.

THE MISÈRE MEX RULE

The Nim-addition rule in misère play is:

> *Provided* one of a and b is 0 or 1,
> *then*
> $a + b = a \overset{*}{+} b$

THE MISÈRE NIM RULE

The Misère Mex Rule shows that

$$\{0, 2, 5\} = \{0\} = 1$$

and

$$\{1, 3, 4\} = \{\} = 0$$

both reduce to Nim-heaps, but positions such as

$$\{2, 3\}$$

which *do* have options but *don't* have options 0 or 1, do not.

The Misère Nim Rule is easily proved by induction. For example,

$$\begin{aligned} 4 + 1 &= \{0 + 1, 1 + 1, 2 + 1, 3 + 1, 4 + 0\} \\ &= \{1, 0, 3, 2, 4\} = 5, \end{aligned}$$

$$\begin{aligned} 5 + 1 &= \{0 + 1, 1 + 1, 2 + 1, 3 + 1, 4 + 1, 5 + 0\} \\ &= \{1, 0, 3, 2, 5, 5\} = 4, \end{aligned}$$

using the Misère Mex Rule.

As Easy as Two and Two?

What do 2 and 2 make? The common man thinks 4, the *normal* Nim player says 0, but the *misère* answer is the $\mathcal{P}$-position

$$2+2=\{2+1,2+0\}=\{3,2\}$$

which can't be simplified any further. The only Nim-heap which is a misère $\mathcal{P}$-position is 1, but

$$1+1 \text{ and } (2+2)+(2+2)$$

have different outcomes, so that certainly we can't simplify $2+2$ to 1.

If addition sums like this had had more satisfactory answers, the misère theory would be easy. Because it isn't, the games grow very complicated and we need our patent collapsing notation to record them.

When several positions have been given names, such as

$$a, b, c, \ldots$$

we shall write

$$a_n \text{ for } a+n, \text{ etc.}$$

and, for example,

$$a_m b_n cd$$

for the game

$$\{a+m, b+n, c, d\}.$$

However, to avoid ambiguity when there's a single option, we'll write

$$a_+ \text{ for } \{a\}.$$

The Misère Form of Grundy's Game

Recall from Chapter 4 that in Grundy's Game the move is to divide any heap into two non-empty heaps of different sizes. Let's begin its analysis in misère play, using the misère nim-addition and mex rules, and writing G_n, for a Grundy heap of size n:

$$\begin{aligned}
\mathrm{G}_1 &= \{\ \ \} = 0\\
\mathrm{G}_2 &= \{\ \ \} = 0,\\
\mathrm{G}_3 &= \{G_2+G_1\} = \{0\} = 1,\\
\mathrm{G}_4 &= \{G_3+G_1\} = \{1\} = 0,\\
\mathrm{G}_5 &= \{G_4+G_1, G_3+G_2\} = \{0,1\} = 2,\\
\mathrm{G}_6 &= \{G_5+G_1, G_4+G_2\} = \{2,0\} = 1,\\
\mathrm{G}_7 &= \{1+0, 2+0, 0+1\} = \{1,2,1\} = 0,\\
\mathrm{G}_8 &= \{0+0, 1+0, 2+1\} = \{0,1,3\} = 2,\\
\mathrm{G}_9 &= \{2+0, 0+0, 1+1, 2+0\} = \{2,0,0,2\} = 1,\\
\mathrm{G}_{10} &= \{1+0, 2+0, 0+1, 1+0\} = \{1,2,1,1\} = 0,\\
\mathrm{G}_{11} &= \{0+0, 1+0, 2+1, 0+0, 1+2\} = \{0,1,3,0,3\} = 2,\\
\mathrm{G}_{12} &= \{2+0, 0+0, 1+1, 2+0, 0+2\} = \{2,0,0,2,2\} = 1.
\end{aligned}$$

But number 13 is unlucky! We find

$$G_{13} = \{1+0, 2+0, 0+1, 1+0, 2+2, 0+1\} = \{1, 2, 1, 1, 2+2, 1\}$$

so that

$$G_{13} = \{2+2, 2, 1\},$$

or, in the collapsed notation,

$$G_{13} = 2_2 21.$$

Since $2+2$ won't reduce to a Nim-heap, we're stuck! The authors don't even know how to play the misère sum of G_{13} with three arbitrary Nim-heaps. From Table 1 you can see that the triples of Nim-heaps x, y, z for which

$$G_{13} + x + y + z$$

is a $\mathcal{P}$-position are quite chaotic. You mustn't expect any magic formula for dealing with such positions.

$x =$ / y	0	1	2	3	4	5	6	7	8	9	10	11	12	13	14	15	16	17	18	19	20	21	22	23
0	1	0	4	5	2	3	8	9	6	7	12	13	10	11	16	17	14	15	20	21	18	19	24	25
1	0	1	5	4	3	2	9	8	7	6	13	12	11	10	17	16	15	14	21	20	19	18	25	24
2	4	5	3	2	0	1	7	6	9	8	11	10	13	12	15	14	17	16	19	18	21	20	23	22
3	5	4	2	3	1	0	6	7	8	9	10	11	12	13	14	15	16	17	18	19	20	21	22	23
4	2	3	0	1	4	5	10	11	15	14	6	7	16	17	9	8	12	13	24	25	22	23	20	21
5	3	2	1	0	5	4	11	10	14	15	7	6	17	16	8	9	13	12	25	24	23	22	21	20
6	8	9	7	6	10	11	3	2	0	1	4	5	14	15	12	13	18	19	16	17	24	25	26	27
7	9	8	6	7	11	10	2	3	1	0	5	4	15	14	13	12	19	18	17	16	25	24	27	26
8	6	7	9	8	15	14	0	1	3	2	16	17	18	19	5	4	10	11	12	13	26	27	29	28
9	7	6	8	9	14	15	1	0	2	3	17	16	19	18	4	5	11	10	13	12	27	26	28	29
10	12	13	11	10	6	7	4	5	16	17	3	2	0	1	18	19	8	9	14	15	29	28	31	30
11	13	12	10	11	7	6	5	4	17	16	2	3	1	0	19	18	9	8	15	14	28	29	30	31
12	10	11	13	12	16	17	14	15	18	19	0	1	3	2	6	7	4	5	8	8	30	31	32	33
13	11	10	12	13	17	16	15	14	19	18	1	0	2	3	7	6	5	4	9	8	31	30	33	32
14	16	17	15	14	9	8	12	13	5	4	18	19	6	7	3	2	0	1	10	11	32	33	34	35
15	17	16	14	15	8	9	13	12	4	5	19	18	7	6	2	3	1	0	11	10	33	32	35	34
16	14	15	17	16	12	13	18	19	10	11	8	9	4	5	0	1	3	2	6	7	34	35	36	37
17	15	14	16	17	13	12	19	18	11	10	9	8	5	4	1	2	2	3	7	6	35	34	37	36
18	20	21	19	18	24	25	16	17	12	13	14	15	8	9	10	11	6	7	3	2	0	1	38	39
19	21	20	18	19	25	24	17	16	13	12	15	14	9	8	11	10	7	6	2	3	1	0	39	38

Table 1. Values of x for Which $G_{13} + x + y + z$ Is a $\mathcal{P}$-Position.

But you probably don't want to play sums of G_{13} with *Nim*-heaps—safter all in Grundy's game you only need play it with other *Grundy* heaps. Even if G_{13} *isn't* a Nim-position, it's still a game; let's call it a and carry on:

$$\mathrm{G}_{14} = \{a+0, 1+0, 2+1, 0+0, 1+2, 2+1\} = a310$$

which reduces to $2 = \{1, 0\}$ since both $a = 2_2 21$ and $3 = 210$ have this as an option. Even though G_{13} wasn't a Nim-heap, G_{14} is! Carrying on still further we find two more Nim-heaps, two games which reduce to a and two new values:

$$\begin{aligned}
\mathrm{G}_{15} &= \mathbf{a20} = 1,\\
\mathrm{G}_{16} &= \mathbf{a_1 2_2 21} = a\\
\mathrm{G}_{17} &= \mathbf{a310} = 2,\\
\mathrm{G}_{18} &= a_2 a20 = b,\\
\mathrm{G}_{19} &= \mathbf{ba_1 2_2 21} = a,\\
\mathrm{G}_{20} &= ba3 = c,\\
\mathrm{G}_{21} &= \mathbf{cb_1} a_2 a20 = b,
\end{aligned}$$

where reversible options are in bold type. From now on all the positions are different. The next is

$$\mathrm{G}_{22} = cba_1 2_2 1 = d,$$

which, however, usually behaves in much the same way as a.

Grundy's Game only reduces to Nim when all the Grundy heaps have sizes

$$1, 2, 3, 4, 5, 6, 7, 8, 9, 10, 11, 12, 14, 15, 17.$$

You can also throw in some heaps of size 13, 16 and 19 by *pretending* that $a + a = 0$ and using the rule:

> If $x, y, z, \ldots$ are Nim-heaps of size at most three, then
>
> $a + x + y + z + \cdots$
>
> is a $\mathcal{P}$-position just if
>
> $x \overset{*}{+} y \overset{*}{+} z \overset{*}{+} \cdots = 1$ or 3
>
> *and* the number of 2- and 3-heaps is exactly
>
> 0 or 3 or 5 or 7 or 9 or ...

If you want to know more, look in the last section of the chapter.

Animals and Their Genus

In later sections we shall show that the Misère Nim strategy extends to a large class of games we shall call **tame**, and that we can say a fair amount about some more games called **restive**, and a little bit about some **restless** games. The **genus** will help to classify these.

Recalling that the normal nim-value $\mathcal{G}^+(G)$ is the unique size of Nim-heap n for which $G+n$ is a $\mathcal{P}$-position in normal play, we now define the **misère nim-value** $\mathcal{G}^-(G)$ to be the unique n which makes $G+n$ a $\mathcal{P}$-position in misère play. You can work out these values by:

$\mathcal{G}^+(G) = 0$ if G has no option, $\mathcal{G}^+(G) = \operatorname{mex} \mathcal{G}^+(G')$ otherwise,	$\mathcal{G}^-(G) = 1$ if G has no option, $\mathcal{G}^-(G) = \operatorname{mex} \mathcal{G}^-(G')$ otherwise,

where G' ranges over all options of G.

Unfortunately $\mathcal{G}^-(G)$ does not enable us to compute $\mathcal{G}^-(G+2)$. So we shall define a more complicated symbol, the **genus** ($\mathcal{G}$-ness)

$$g^{\gamma_0\gamma_1\gamma_2\gamma_3\ldots}$$

of G, where

$$g = \mathcal{G}^+(G),$$

$$\gamma_0 = \mathcal{G}^-(G), \quad \gamma_1 = \mathcal{G}^-(G+2), \quad \gamma_2 = \mathcal{G}^-(G+2+2), \quad \gamma_3 = \mathcal{G}^-(G+2+2+2), \ldots.$$

We shall abbreviate this symbol in various ways.

g denotes a Nim-heap of size g.
g^γ denotes a tame or restive game.
(In these cases the full genus can be recovered from Table 2).
$g^{\alpha\beta\ldots\lambda\mu}$ abbreviates the genus $g^{\alpha\beta\ldots\lambda\mu\lambda\mu\lambda\mu\ldots}$

It turns out that *every* genus ends up alternating between two numbers like this.

Nim-heaps	and	Tame Games	Restive Games			
g	g^γ	Genus	g^γ	Genus	g^γ	Genus
0	0^1	$0^{12020\ldots}$				
1	1^0	$1^{03131\ldots}$				
	0^0	$0^{02020\ldots}$				
	1^1	$1^{13131\ldots}$				
2	2^2	$2^{20202\ldots}$	0^2	$0^{20202\ldots}$	1^2	$1^{20202\ldots}$
3	3^3	$3^{31313\ldots}$	0^3	$0^{31313\ldots}$	1^3	$1^{31313\ldots}$
4	4^4	$4^{46464\ldots}$	0^4	$0^{42020\ldots}$	1^4	$1^{43131\ldots}$
5	5^5	$5^{57575\ldots}$	0^5	$0^{52020\ldots}$	1^5	$1^{53131\ldots}$
6	6^6	$6^{64646\ldots}$	0^6	$0^{62020\ldots}$	1^6	$1^{63131\ldots}$
7	7^7	$7^{75757\ldots}$	0^7	$0^{72020\ldots}$	1^7	$1^{73131\ldots}$

Table 2. Abbreviations for the Genus of Certain Games

For Grundy's Game the genus sequence with these abbreviations is

$$.0\ 0\ 1\ 0\ 2\ 1\ 0\ 2\ 1\ 0\ 2\ 1\ 3^{1431}\ 2\ 1\ 3^{1431}\ 2\ 4^{0564}\ 3^{1431}\ 0^{20}\ 4^{0564}\ 3^{1431} \ldots$$

and for Kayles,

K_0	K_1	K_2	K_3	K_4	K_5	K_6	K_7	K_8	K_9	K_{10}	K_{11}	K_{12}	K_{13}	K_{14}	K_{15}	K_{16}	K_{17}	...
0.	1	2	3	1	4^{146}	3	2^2	1^1	4^{046}	2^2	6^{46}	4^{046}	1^1	2^2	7^{57}	1^{13}	4^{64}	...

The single superscript form is reserved for tame or restive games. The games K_{13} and K_{16} have the same genus

$$1^{1313\ldots}$$

but our displayed sequence abbreviates them differently since only the former is tame.

What Can We With the Genus?

The genus is a very useful tool in misère play calculations, because, as we shall gradually come to understand:

1. We can find the misère outcome of a single game. The outcome is $\mathcal{P}$ if and only if the first superscript is 0.
2. If each of our games is *tame* and of known genus, we can work out the genus—and therefore the outcome—of their sum, which is also tame.
3. If $a, b, c, \ldots$ are Nim-heaps of sizes 0,1,2,3, then, from the genus of an arbitrary G, we can find the genus and outcome of

   $$G + a + b + c + \ldots.$$

4. We can allow *one* of these Nim-heaps to have size 4 or more and still compute the outcome.
5. If $a, b, c, \ldots$ are *any* Nim-heaps and R is *restive* with given genus, we can find the genus and outcome of

   $$R + a + b + c + \ldots.$$

6. For many restive games, R, our Noah's Ark Theorem shows that

   $$R + R, \qquad R + R + R, \ldots$$

   are tame and of known genus and outcome.
7. And there are many more *half-tame* games, H, for which $H + H$ is tame; again of known genus and outcome.
8. For a suitably restrained *restless* game, R, an even number of copies of R will not affect the genus or outcome in suitable circumstances, again by the Noah's Ark Theorem.

Firm, Fickle and Tame

In Nim only the following combinations arise:

$$\begin{array}{ll} 0^1, 1^0 & \text{(the \textbf{fickle units})} \\ 0^0, 1^1 & \text{(the \textbf{firm units})} \\ 2^2, 3^3, 4^4, 5^5, \ldots & \text{(the \textbf{big firms})} \end{array}$$

and they can be combined according to the following rule, which works for all tame games.

> If *any* component is firm, so is the sum, and
>
> $$a^\alpha + b^\beta + \cdots = (a \overset{*}{+} b \overset{*}{+} \cdots)^{\alpha \overset{*}{+} \beta \overset{*}{+} \cdots}.$$
>
> If *all* components are fickle, so is the sum, and
>
> $$a^\alpha + b^\beta + \cdots = (a \overset{*}{+} b \overset{*}{+} \cdots)^{1 \overset{*}{+} \alpha \overset{*}{+} \beta \overset{*}{+} \cdots}.$$

COMBINING TAME GAMES

Thus

$$0^1 + 2^2 + 7^7 = 5^7,$$

having two firm components, must be firm of genus 5^5, but

$$0^1 + 1^0 + 1^0 = 0^?,$$

a completely fickle sum, has genus 0^1.

Let's look at the Kayles position

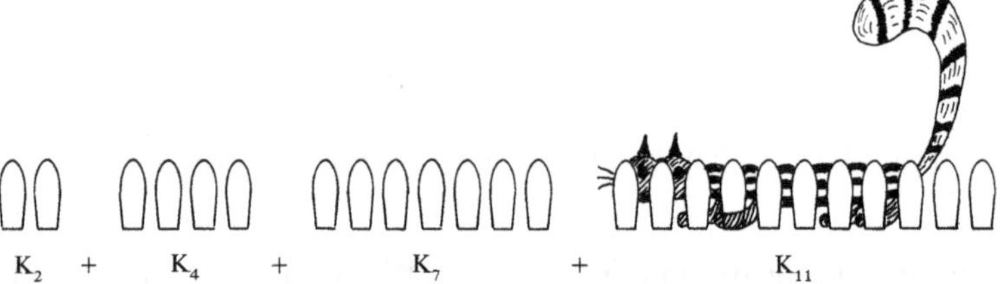

Our sequence indicated that K_2, K_4 and K_7 are tame; in fact K_2 and K_4 are *very* tame, being Nim-heaps. The three tame components together make a tame game of genus 1^1,

$$2^2 + 1^0 + 2^2 = 1^1.$$

But K_{11} is a really wild animal of genus $6^{4646\ldots}$. What shall we do? Luckily we spot K_{11}'s tame option

$$K_3 + K_7$$

of genus $3^3 + 2^2 = 1^1$,

leading to

$$K_2 + K_4 + K_7 + K_3 + K_7,$$

whose genus

$$2 + 1 + 2^2 + 3 + 2^2 = 0^0$$

shows it to be a $\mathcal{P}$-position.

Which Animals are Tame...

Roughly speaking, those we can pretend are Nim-positions. In this book we have managed to tame many more animals than were regarded as tame in ONAG (**hereditarily tame**). But see the end of the Extras for the even larger family of **tameable games**.

For a game G to be **tame** its nim-values g^γ must certainly form one of the **tame pairs**

$$0^0, 1^1, 2^2, 3^3, 4^4, \ldots \text{ or } 0^1, 1^0$$

that arise for Nim-positions. There is no further condition if all the options of G are tame, but *wild* options of G are also allowed provided they aren't needed to determine the nim-values of G, and provided they each have **reverting** moves to *tame* games with nim-values $g^?$ and $?^\gamma$.

More precisely, if the tame options have genus

$$a^\alpha, b^\beta, \ldots$$

we must have

$$g = \operatorname{mex}(a, b, \ldots)$$
$$\gamma = \operatorname{mex}(\alpha, \beta, \ldots)$$

and, from any wild option, there must be moves to two (possibly equal) tame games with genus

$$g^? \text{ and } ?^\gamma.$$

For example, the Kayles position

$$\begin{aligned} \mathrm{K}_7 &= \{\mathrm{K}_6, \mathrm{K}_5 + \mathrm{K}_1, \mathrm{K}_4 + \mathrm{K}_2, \mathrm{K}_3 + \mathrm{K}_3, \mathrm{K}_5, \mathrm{K}_4 + \mathrm{K}_1, \mathrm{K}_3 + \mathrm{K}_2\} \\ &= \{3, \quad \mathrm{K}_5 + 1, \quad 1 + 2, \quad 3 + 3, \quad \mathrm{K}_5, \quad 1 + 1, \quad 3 + 2\} \end{aligned}$$

has tame options

$$3^3, \qquad 3^3, \qquad 0^0, \qquad 0^1, \qquad 1^1$$

which suffice to determine the genus 2^2. The two wild options have reverting moves to tame games with this genus.

$$\mathrm{K}_5 + 1 \to \mathrm{K}_3 + 1, \text{ genus } 2^2,$$
$$\mathrm{K}_5 \to \mathrm{K}_3 + \mathrm{K}_1, \text{ genus } 2^2,$$

and so K_7 is itself tame of genus 2^2.

To win a sum of tame games make sure that after each of your moves every component is tame and the total genus is

0^0 if *any* component is *firm*,
1^0 if *every* component is *fickle*.

If your opponent ever plays a wild card there will always be a suitable reverting move for you to respond with. Otherwise the strategy is exactly as in Misère Nim.

... and Which are Restive?

Once again we have a wider class than was treated in ONAG. For G to be **restive**, its nimvalues g^γ have to be one of the **restive pairs** in which g is 0 or 1 and γ is 2 or more:

$$0^2, 0^3, 0^4, 0^5, \ldots \text{ or } 1^2, 1^3, 1^4, 1^5, \ldots.$$

There is no further condition if all the options of G are tame, but wild options are also allowed provided they are not needed to determine the nim-values of G, and each has *reverting* moves to *tame* or *restive* games of genus

$$g^{?} \text{ and } ?^{\gamma}$$

where each ? must be one of

$$0, 1, \gamma, \gamma \overset{*}{+} 1.$$

More precisely, if the tame options have genus

$$a^{\alpha}, b^{\beta}, \ldots$$

we must have

$$g = \operatorname{mex}(a, b, \ldots) = 0 \text{ or } 1,$$

$$\gamma = \operatorname{mex}(a, \beta, \ldots) \geq 2,$$

and, from any wild option there must be moves to two (possibly equal) tame or restive games of genus

$$g^{?} \text{ and } ?^{\gamma} \text{ (each } ? = 0, 1, \gamma, \text{ or } \gamma \overset{*}{+} 1).$$

The situation can get quite awkward when you put restive animals together. We wouldn't even like to have to predict the outcome when a restive animal is put among certain tame ones but at least they behave well with Nim-positions.

Question: Suppose R is a restive game of genus g^{γ} and m, n, ... are Nim-heaps; when is $R + m + n + \cdots$ a $\mathcal{P}$-position ? *Answer:* This happens when $m \overset{*}{+} n \overset{*}{+} \cdots = \gamma$, if each of m, n, ... is 0, 1, γ or $\gamma \overset{*}{+} 1$; and when $m \overset{*}{+} n \overset{*}{+} \cdots = g$, otherwise.

RESTIVE GAMES ARE AMBIVALENT NIM-HEAPS

Moreover

If $0, 1 < N < \gamma, \gamma \overset{*}{+} 1$ then $R + n$ is tame of genus $(g \overset{*}{+} n)^{\gamma \overset{*}{+} n}$

THE INTERMEDIATE VALUE THEOREM

Some Tame Animals in the Good Child's Zoo

Misère Wyt Queens

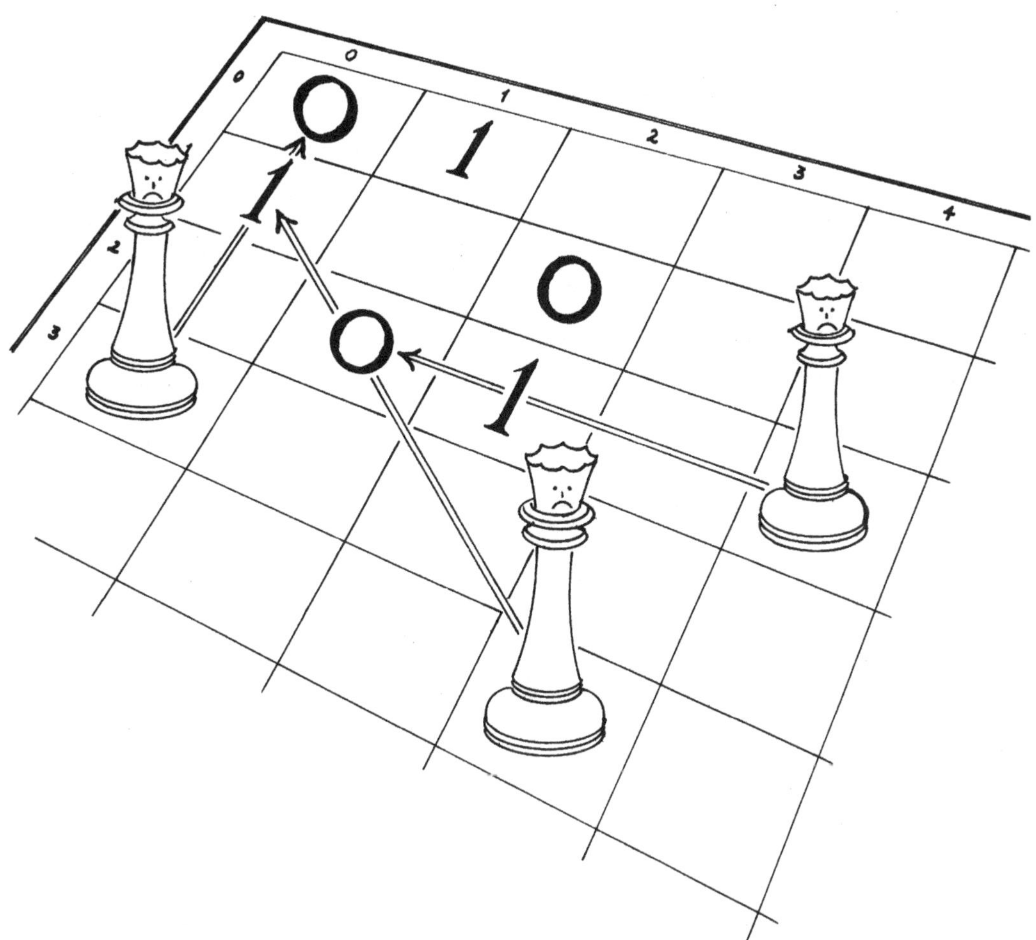

Figure 5. Why Wyt Queens Are Tame.

We met the normal play version of Wyt Queens in Chapter 3, and you'll find by trial that the misère version has a genus table which appears to be entirely tame (Table 3). Figure 5 shows how every Queen outside the leading 3×3 square which can "see" a 0 or 1 in this square can also see the other, and using this we can prove that every entry outside this square is both firm and tame. Welter's Game and Moore's Nim_k are also tame (see Chapter 15). Yamasaki independently gives the misère analysis of all these games. Wythoff's Game is also known as Chinese Nim or Tsyan-shizi, and Welter's Game as Sato's Maya Game.

	0	1	2	3	4	5	6	7	8	9	10	11	12	13	14	15
0	0	1	2	3	4	5	6	7	8	9	10	11	12	13	14	
1	1	2	0	4	5	3	7	8	6	10	11	9	13	14	12	
2	2	0	1	5	3	4	8	6	7	11	9	10	14	12	13	
3	3	4	5	6	2	0^0	1^1	9^9	10^{10}	12^{12}	8^8	7^7	15^{15}	11^{11}	16^{16}	
4	4	5	3	2	7	6	9^9	0^0	1^1	8^8	13^{13}	12^{12}	11^{11}	16^{16}	15^{15}	
5	5	3	4	0^0	6	8^8	10^{10}	1^1	2^2	7^7	12^{12}	14^{14}	9^9	15^{15}	17^{17}	
6	6	7	8	1^1	9^9	10^{10}	3	4	5^5	13^{13}	0^0	2^2	16^{16}	17^{17}	18^{18}	
7	7	8	6	9^9	0^0	1^1	4	5	3	14^{14}	15^{15}	13^{13}	17^{17}	2^2	10^{10}	
8	8	6	7	10^{10}	1^1	2^2	5^5	3	4	15^{15}	16^{16}	17^{17}	18^{18}	0^0	9^9	
9	9	10	11	12^{12}	8^8	7^7	13^{13}	14^{14}	15^{15}	16^{16}	17^{17}	6^6	19^{19}	5^5	1^1	
10	10	11	9	8^8	13^{13}	12^{12}	0^0	15^{15}	16^{16}	17^{17}	14^{14}	18^{18}	7^7	6^6	2^2	
11	11	9	10	7^7	12^{12}	14^{14}	2^2	13^{13}	17^{17}	6^6	18^{18}	15^{15}	8^8	19^{19}	20^{20}	
12	12	13	14	15^{15}	11^{11}	9^9	16^{16}	17^{17}	18^{18}	19^{19}	7^7	8^8	10^{10}	20^{20}	21^{21}	
13	13	14	12	11^{11}	16^{16}	15^{15}	17^{17}	2^2	0^0	5^5	6^6	19^{19}	20^{20}	9^9	7^7	
14	14	12	13	16^{16}	15^{15}	17^{17}	18^{18}	10^{10}	9^9	1^1	2^2	20^{20}	21^{21}	7^7	11^{11}	

Table 3. The Genus of Wyt Queens.

Jelly Beans and Lemon Drops

The **Jelly Bean Game, ·52**, may be played with rows of jelly beans. We may take away 1 bean, provided it is strictly internal to its row, or forms the whole of its row; or 2 adjacent beans in precisely the opposite case, when they are at the end of the row but are *not* the whole row. Using J_n for a row of n jelly beans we find

$$\mathrm{J}_1 = 1, \quad \mathrm{J}_2 = 0, \quad \mathrm{J}_3 = 2, \quad \mathrm{J}_4 = 2, \quad \mathrm{J}_5 = 1,$$
$$\mathrm{J}_6 = \{2, 3\} = 2_2 \text{ (genus } 0^0), \quad \mathrm{J}_7 = 3,$$
$$\mathrm{J}_8 = 3_2 2_2 1 \text{ (genus } 2^2), \quad \mathrm{J}_9 = 2_2 32(1^1).$$

Continuing the calculation,

$$\begin{array}{llllllll} J_{10} = & \{J_8+J_1, & J_7+J_2, & J_6+J_3, & J_5+J_4, & J_8\}, & \\ \text{genus} & \{3^3, & 3^3, & 2^2, & 3^3, & 2^2\} = 0^0, & \\ J_{11} = & \{J_9+J_1, & J_8+J_2, & J_7+J_3, & J_6+J_4, & J_5+J_5, & J_9\}, \\ \text{genus} & \{0^0, & 2^2, & 1^1, & 2^2, & 0^1, & 1^1\} = 3^3 \end{array}$$

This is the last case which has a fickle option, $J_5 + J_5 = 0^1$. From now on every J_n will be both firm and tame since all its options are, and the genus sequence

$$.1\ \ 0\ \ 2\ \ 2\ \ 1;\ \ 0^0\ \ 3\ \ 2^2\ \ 1^1\ \ 0^0\ \ 3^3\ \ 2^2\ \ 1^1\ \ 0^0\ \ 3^3\ldots$$

has period 4 after the semi-colon.

When we also allow 2 adjacent beans to be taken from *inside* a longer row we have the game of **Lemon Drops, ·56**, with typical position L_n. The complete analysis of **·56** for normal play still eludes us, despite computations for rows of up to 50000 drops. But at least it's no harder to play the misère version, since after $L_6 = 1$ every position is firm and tame:

$$.1\ \ 0\ \ 2\ \ 2\ \ 4\ \ 1\ \ 1^1\ \ 3\ \ 2\ \ 4^4\ \ 4^4\ \ 6^6\ \ 6^6\ \ 2^2\ \ 1^1\ \ 1^1\ \ 7^7\ \ 6^6\ \ 8^8\ \ 4^4\ \ 1^1\ \ 1^1\ldots.$$

Stalking Adders and Taking Squares

For **Stalking, ·31**, the reduced forms are

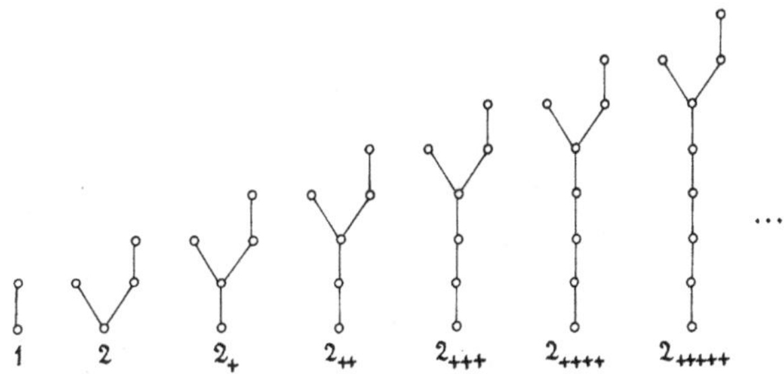

showing that the game is tame with genus sequence

$$.1\ \ 2\ \ 0^0\ \ 1^1\ \ 0^0\ \ 1^1\ \ 0^0\ \ 1^1\ \ 0^0\ldots$$

while for **Adders, ·73**, the reduced forms are

$$1\ \ 2\ \ 3\ \ 2_2\ \ 3_2\ \ 2_{22}\ \ 3_{22}\ \ 2_{222}\ \ 3_{222}\ \ 2_{2222}\ \ 3_{2222}\ldots$$

again tame, with genus sequence

$$.1\ \ 2\ \ 3\ \ 0^0\ \ 1^1\ \ 2^2\ \ 3^3\ \ 0^0\ \ 1^1\ \ 2^2\ \ 3^3\ldots$$

Of course, 2_{++} and 3_{22} are merely the collapsed notations for

$$(2_+)_+ = \{\{2\}\} \text{ and } (3_2)_2 = 3+2+2.$$

Take-A-Square is another heap game. Each player in turn chooses a heap and takes

1 or 4 or 9 or 16 or ...

beans from it. If S_n is a heap of size n,

$$\mathrm{S}_n = \{\mathrm{S}_{n-1}, \mathrm{S}_{n-4}, \mathrm{S}_{n-9}, ...\}$$

and the first few S_n reduce to Nim-heaps:

n	0	1	2	3	4	5	6	7	8	9	10	11	12	13	14	15	16	17	18	19	20	21	22	23	24	25	...
S_n	0	1	0	1	2	0	1	0	1	2	0	1	0	1	2	0	1	0	1	2	0	1	0	1	2	3	....

Although 2 and 3 appear we'll never get any positions like $\{2,3\}$ which don't reduce to Nim-heaps. In the Extras we'll give T.S. Ferguson's proof that *all* subtraction games reduce to Nim, and so are tame. Yamasaki has independently reached the same conclusion. His terms "flat" and "projective" both imply our "tame".

"But What if They're Wild?" Asks the Bad Child

She'll need to know how to compute the genus:

For a non-empty game

$$G = \{A, B, \ldots\}$$

where

A has genus $a^{\alpha_0\alpha_1\alpha_2\ldots}$

B has genus $b^{\beta_0\beta_1\beta_2\ldots}$

....

the genus

$$g^{\gamma_0\gamma_1\gamma_2\ldots}$$

is found from

$$g = \operatorname{mex}(a, b, \ldots)$$
$$\gamma_0 = \operatorname{mex}(\alpha_0, \beta_0, \ldots)$$
$$\gamma_1 = \operatorname{mex}(\gamma_0, \gamma_0 \overset{*}{+} 1, \alpha_1, \beta_1, \ldots)$$
$$\gamma_2 = \operatorname{mex}(\gamma_1, \gamma_1 \overset{*}{+} 1, \alpha_2, \beta_2, \ldots)$$

....

$$\gamma_{n+1} = \operatorname{mex}(\gamma_n, \gamma_n \overset{*}{+} 1, \alpha_{n+1}, \beta_{n+1}, \ldots)$$

and how to add small Nim-heaps:

$$
\begin{array}{lll}
\text{If} & G & \text{has genus } g^{\gamma_0\gamma_1\gamma_2\gamma_3\cdots} \\
\text{then} & G+1 & \text{has genus } (g \overset{*}{\ddagger} 1)^{\delta_0\delta_1\delta_2\delta_3\cdots} \\
 & G+2 & \text{has genus } (g \overset{*}{\ddagger} 2)^{\gamma_1\gamma_2\gamma_3\cdots} \\
 & G+3 & \text{has genus } (g \overset{*}{\ddagger} 3)^{\delta_1\delta_2\delta_3\cdots} \\
 & G+2+2 & \text{has genus } g^{\gamma_2\gamma_3\cdots} \\
 & G+3+2 & \text{has genus } (g \overset{*}{\ddagger} 1)^{\delta_2\delta_3\cdots} \\
 & \cdots\cdots\cdots\cdots\cdots\cdots & \\
\multicolumn{3}{l}{\text{where } \delta_0 = \gamma_0 \overset{*}{\ddagger} 1,\ \delta_1 = \gamma_1 \overset{*}{\ddagger} 1,\ \delta_2 = \gamma_2 \overset{*}{\ddagger} 1,\ \ldots}
\end{array}
$$

Here's how we find the genus for four of Grundy's wild animals ($a = 2_2 21$, $b = a_2 a 20$, $c = ba3$, $d = cba_1 2_2 1$):

						c	$0^{2020\ldots}$
		a_2	$1^{4313\ldots}$			b	$4^{0564\ldots}$
2_2	$0^{0202\ldots}$	a	$3^{1431\ldots}$	b	$4^{0564\ldots}$	a_1	$2^{0520\ldots}$
2	$2^{2020\ldots}$	2	$2^{2020\ldots}$	a	$3^{1431\ldots}$	2_2	$0^{0202\ldots}$
1	$1^{0313\ldots}$	0	$0^{1202\ldots}$	3	$3^{3131\ldots}$	1	$1^{0313\ldots}$
a	$3^{1431\ldots}$	b	$4^{0564\ldots}$	c	$0^{2020\ldots}$	d	$3^{1431\ldots}$

Because Grundy's Game was extensively analyzed in ONAG, we'll leave further analysis until later and use Kayles as our next example.

Misère Kayles

For the tame animals in our good child's zoo we didn't need reverting moves since no wild game crossed our path. Although several tame games arise in Kayles (see Chapter 4), wild game's abounding and we'll need all our resources to tackle it. Table 4 pushes the genus analysis to heaps of size 20, and with increasing difficulty we find the next few terms:

K_{21}	K_{22}	K_{23}	K_{24}	K_{25}	...
4^{64}	6^{46}	7^{57}	4^{64}	1^{731}	...

The options of each game are found on the preceding two lines of the table. Thus

$$K_{10} = \delta = \{\gamma, \beta+2, \beta+1, \alpha+3, \alpha+1, 3+2, 2+2, 0\},$$

whose tame options: ↓ 0^0, 3^3, ↓ ↓ 1^1, 0^0, 0^1 yield 2^2,

while others revert to: 2, 2, 2,

proving that K_{10} is tame with genus 2^2. In normal Kayles, no single row other than 0 can be a $\mathcal{P}$-position (take out the middle 1 or 2 pins), but the table shows that in misère play,

$$K_1, K_4, K_9, K_{12}, K_{20}$$

are $\mathcal{P}$-positions. It is not hard to justify enough $\mathcal{P}$-positions:

n	K_n	$K_{n-1}+K_1$	$K_{n-2}+K_2$	$K_{n-3}+K_3$	$K_{n-4}+K_4$	$K_{n-5}+K_5$	$K_{n-6}+K_6$	$K_{n-7}+K_7$	$K_{n-8}+K_8$	$K_{n-9}+K_9$
0	0									
1	1									
2	2	0								
3	3	3								
4	1	2	$0^0(2_2)$							
5	$4^{146}(\alpha)$	0	$1^1(3_2)$							
6	3	$5^{057}(\alpha_1)$	3	$0^0(2_2)$						
7	$2^2(\beta)$	2	$6^{46}(\alpha_2)$	2						
8	$1^1(3_2)$	$3^3(\beta_1)$	$1^1(3_2)$	$7^{57}(\alpha_3)$	0					
9	$4^{046}(\gamma)$	0^0	0^0	0^0	5^{057}					
10	$2^2(\delta)$	5^{157}	3^3	1^1	2	0^{120}				
11	6^{46}	3^3	6^{46}	2^2	3^3	7^{57}				
12	4^{046}	7^{57}	0^0	7^{57}	0^0	6^{46}	0^0			
13	1^1	5^{157}	4^{64}	1^1	5^{157}	5^{75}	1^1			
14	2^2	0^0	6^{46}	5^{75}	3^3	0^{02}	2^2	0^0		
15	7^{57}	3^3	3^3	7^{57}	7^{57}	6^{46}	7^{57}	3^3		
16	1^{13}	6^{46}	0^0	2^2	5^{157}	2^{20}	1^1	6^{46}	0^0	
17	4^{64}	0^{02}	5^{75}	1^1	0^0	0^{02}	5^{75}	0^0	5^{75}	
18	3^{31}	5^{75}	3^{31}	4^{64}	3^3	5^{75}	7^{57}	4^{64}	3^3	0^{120}
19	2^{20}	2^{20}	6^{46}	2^{20}	6^{46}	6^{46}	2^2	6^{46}	7^{57}	6^{46}
20	1^{031}									

$\alpha = 2_2321$
$\beta = \alpha_1\alpha 3_2 2_2 30$
$\gamma = \beta_1\beta\alpha_3\alpha_2 3_2 20$
$\delta = \gamma\beta_2\beta_1\alpha_3\alpha_1 3_2 2_2 0$

Table 4. Genus Analysis of Misère Kayles.

$$K_{20}, \quad K_{11}+K_{11}, \quad K_8+K_{16}, \quad K_{13}+K_{13}, \quad K_{14}+K_{14}, \quad K_{15}+K_{15}, \quad K_{16}+K_{16},$$

to provide good replies to

$$K_{21},K_{22}, \quad K_{23},K_{24}, \quad K_{25},K_{26}, \quad K_{27},K_{28}, \quad K_{29},K_{30}, \quad K_{31},K_{32}, \quad K_{33},K_{34}.$$

Since the first edition, William Sibert has found a complete analysis of Misère Kayles — see the Extras.

The Noah's Ark Theorem

Any wild game that's only just gone wild, i. e. has only tame options, must have among its options one, but not both, of the two kinds $(0^1, 1^0)$ of fickle unit and one, but not both, of the two kinds $(0^0, 1^1)$ of firm unit. If the options include both kinds of 0 or both kinds of 1:

$$0^1, 0^0, a^a, b^b, \ldots \text{ or } 1^0, 1^1, a^a, b^b, \ldots \qquad (2 \le a, b, \ldots)$$

the game is restive. If one of each:

$$1^0, 0^0, a^a, b^b, \ldots \text{ or } 0^1, 1^1, a^a, b^b, \ldots \qquad (2 \le a, b, \ldots)$$

we call it **restless**.

Usually two copies of a restive game make a firm zero, while two copies of a restless game can be treated as a fickle zero. So (look at Fig. 6) we can allow some restive and restless animals into our ark of tame creatures provided that they come in pairs and are suitably restrained:

Figure 6. The Noah's Ark Theorem

Suppose that

$$\begin{array}{ll} T_1,\ T_2,\ \ldots & \text{are tame,} \\ R_1,\ R_2,\ \ldots & \text{are restive, and the game} \\ R & \text{is restless of genus } g^{\gamma\cdots} \end{array}$$

and that R_1, R_2, ..., R have only just gone wild.
Then you can find the outcome of

$$T_1 + T_2 + \cdots + (R_1 + R_1) + (R_2 + R_2) + \cdots + (R + R) + (R + R) + \cdots$$

by noting that each pair

$$R_1 + R_1, \quad R_2 + R_2, \quad \ldots$$

is a tame game of genus 0^0 and *neglecting* each pair

$$R + R$$

provided that

when there is any pair $R + R$,
(i) the only fickle tame positions of $T_1, T_2, \ldots, R_1, R_2, \ldots, R$ are 0 and 1, and
(ii) for each option a^a $(a \geq 2)$ of R,
either a^a has an option $\gamma \overset{*}{\ddagger} 1$,
or R has an option $(a \overset{*}{\ddagger} 1)^{a \overset{*}{\ddagger} 1}$

THE NOAH'S ARK THEOREM

The complicated sounding conditions are usually satisfied automatically, because the options a^a in condition (ii) are often Nim-heaps a, and we rarely see a fickle tame position other than 0 or 1.

The strategy showing why $R + R$ can be neglected is given in the Extras. It also proves that when (i) and (ii) are satisfied

$$T_1 + T_2 + \ldots + (R_1 + R_1) + (R_2 + R_2) + \ldots + (R + R) + \ldots + (R + R) + R$$

has the same outcome as

$$T_1 + T_2 + \ldots + (R_1 + R_1) + (R_2 + R_2) + \ldots + R.$$

In particular, all even multiples of R have genus

$$0^{1202\ldots}$$

and all odd ones have the same genus as R.

Thus Grundy's Game first goes wild at $G_{13} = 2_2 21$, which is restless of genus 3^{1431}, and so the sum of any odd number of copies of G_{13} will have the same genus, while the sum of any even number has the same genus 0^{120} as 0. Similar remarks hold for $K_5 = 2_2 321$ of genus 4^{146}, in Kayles.

To show that $R_1 + R_1$ is tame of genus 0^0, we exhibit a reverting move to 0^0 from *every* option:

$$R_1 + R_1 \begin{cases} \nearrow R_1 + n^n \longrightarrow n^n + n^n = 0^0, \\ \searrow \begin{cases} R_1 + 0^1 \longrightarrow 0^0 + 0^1 = 0^0, \\ \text{or} \\ R_1 + 1^0 \longrightarrow 1^1 + 1^0 = 0^0. \end{cases} \end{cases}$$

Moreover if r^ρ is the genus of R_1, then $R_1 + R_1 + R_1$ is tame of genus r^r, because all options are tame and firm:

$$R_1 + R_1 + R_1 \begin{cases} \nearrow R_1 + R_1 + n^n = n^n, \\ \searrow \begin{cases} R_1 + R_1 + 0^1 = 0^0, \\ \text{or} \\ R_1 + R_1 + 1^0 = 1^1. \end{cases} \end{cases}$$

This means that we can allow some pairs of restive animals to bring their children into the ark.

The Half-Tame Theorem

We call a wild animal H **half-tame** if H + H is tame of genus 0^0. The argument we used to prove that $R_1 + R_1$ is tame generalizes to show that:

> A wild game is half tame
> *provided that*
> all options of H are tame or half-tame,
> *and*
> if H has an option 0^1, it has an option 0^0,
> if H has an option 1^0, it has an option 1^1,

THE HALF-TAME THEOREM

It is not actually necessary for the truth of the Noah's Ark Theorem that the games of each restive or restless pair be identical. It will suffice if they are merely of the same **species**. The following operations will not change the species:

(a) replacing a tame option by another tame game of the same genus (observing conditions (i) and (ii) of the Noah's Ark Theorem in the restless case),

(b) adding a new option from which there is a reverting move to a game already known to be of the same species.

Guiles

Guiles is the octal game **·15** played with rows Y_n of n beans from which 2 adjacent beans may be removed provided neither is the end of a row and a complete row that contains 1 or 2 beans may also be removed. The genus sequence is remarkable for its repetitions:

$$0.110112212211 0 1^4 1^4 2 2 1^2 2 2^{1420} 1^4 1^4 0^1 1^4 1^4 2 2^{1420} 1^2 2^{1420} 2^{1420} 1^1 1^{631} \ldots .$$

All the occurrences of 1^4 refer to the same restive game

$$\mathrm{Y}_{14} = \mathrm{Y}_{15} = \mathrm{Y}_{21} = \mathrm{Y}_{22} = \mathrm{Y}_{24} = \mathrm{Y}_{25} = 2_2 320 = z, \text{say},$$

which arises in many other contexts (**·57**, **·72**, **·75** and **4 · 7**). In terms of this

$\mathrm{Y}_{18} = z_1 2_2 0;$	*restive,*	genus 1^2,
$\mathrm{Y}_{20} = z_1 z 2_2 31;$	*wild,*	genus 2^{1420},
$\mathrm{Y}_{23} = \{\mathrm{Y}_{20}+1, \mathrm{Y}_{18}, z+2, 3, 1\};$	*tame*	genus 0^1.

Y_{19} and Y_{26} both reduce to the Nim-heap of size 2!

Why is $\mathrm{Y}_{28} = \{\mathrm{Y}_{23}, \mathrm{Y}_{20}+2, \mathrm{Y}_{18}+1, z+1, 2+2, 0\}$ restive? Its tame options

$$\mathrm{Y}_{23}, 2+2, 0$$

with values

$$0^1, 0^0, 0^1$$

suffice to compute the value 1^2.

From $\mathrm{Y}_{20}+2$	there are reverting moves to $3+2(\mathbf{1}^1)$ and $2+2+2(2^2)$.
From $\mathrm{Y}_{18}+1$	there is a reverting move to $\mathrm{Y}_{18}(\mathbf{1}^2)$.
From $z+1$	there are reverting moves to $0+1(\mathbf{1}^0)$ and $3+1(2^2)$.

Y_{31} is tame of genus 1^1. It has all the options

$$2+2, 3, 2$$

of $3+2$, and from every other option except $z+z$ (which is tame of genus 0^0 by the Noah's Ark Theorem) there is a reverting move to $3+2$, so that Y_{31} is "nearly equal to" $3+2$.

Dividing Rulers

Dividing Rulers is the name for our next game because the positions (Fig. 7) repeat like those of the Ruler Game (Fig. 7 in Chapter 14). The move is to split any heap into two smaller heaps or to halve the size of any even heap. Writing R_n for a heap of n we have, if $n = 2^k d$ and d is odd,

$$\mathrm{R}_n = \mathrm{H}_k, \quad \text{where} \quad \mathrm{H}_0 = 0, \quad \mathrm{H}_1 = 1, \quad \mathrm{H}_2 = 2, \quad \mathrm{H}_3 = 2_2 20,$$

and in general

$$\mathrm{H}_{k+1} = \{\mathrm{H}_0 + \mathrm{H}_0, \mathrm{H}_1 + \mathrm{H}_1, \mathrm{H}_2 + \mathrm{H}_2, \ldots, \mathrm{H}_k + \mathrm{H}_k, \mathrm{H}_k\}.$$

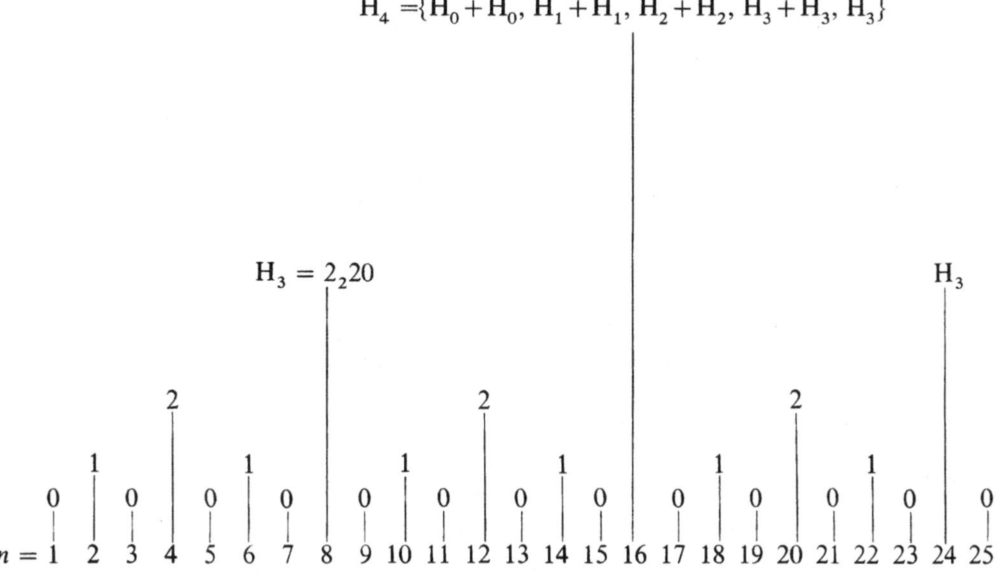

Figure 7. Positions in Dividing Rulers.

This is because the options of R_n have the forms

$$R_{2^{k-1}d} \qquad R_{2^j a} + R_{2^j b} \qquad R_{2^k a} + R_{2^l b}$$

where a, b are odd and $j < k < l$. By induction these are

$$H_{k-1} \qquad H_j + H_j \qquad H_k + H_l.$$

The first two cases are the options defining H_k, and from the third there is a reverting move to $H_k + H_0 + H_0 = H_k$.

Because H_3 is restive and only just wild we know the genus of all its multiples:

	H_3	H_3+H_3	$H_3+H_3+H_3$	$H_3+H_3+H_3+H_3$	$H_3+H_3+H_3+H_3+H_3$	...
genus	1^3	0^0	1^1	0^0	1^1...	

and we can therefore predict the outcome of any position in which no heap has size divisible by 16. Also, because $H_3, H_4, \ldots$ are half-tame it is easy to find their genus:

$$H_3, H_5, H_7, \ldots \text{ have genus } 1^{31},$$
$$H_4, H_6, H_8, \ldots \text{ have genus } 2^{20}.$$

But positions where they don't occur in even numbers can be more complicated; for example

$$H_3 + H_4 \text{ has genus } 3^{46}$$

but

$$H_3 + H_3 + H_3 + H_4 \text{ has genus } 3^{31}.$$

There is a variant in which the division of a heap into two *equal* heaps is not allowed. Using V_n for a heap of size n in the variant, we find that some of the theory still goes through:

$$\mathrm{V}_d = 0, \mathrm{V}_{2d} = 1, \mathrm{V}_{4d} = 2, \text{ if } d \text{ is odd,}$$

but $\mathrm{V}_8, \mathrm{V}_{16}, \mathrm{V}_{24}, \ldots$ are all distinct:

n	8	16	24	32	40	48	56	64	72	80	88	96	104	...
V_n	1	2	z	y	x	w	v	u	t	s	r	q	p	...
genus	1	2	1^4	1^3	1^4	2^{20}	1^4	2^{20}	1^4	2^{20}	1^4	1^{31}	1^4	...

In fact $z = 2_2320$ is the restive game we met in Guiles and the alternate terms thereafter, $x, v, t, \ldots$ are restive of the same species. We show the tame options and reverting moves for the case

	$t = \{u+1,$	$v+2,$	$w+z,$	$x+y,$	$2+2,$	$0,$	$2\}$
tame options:	↓	$3^3,$	↓	↓	$0^0,$	$0^1,$	$2^2,$
reverting moves:	$(v+1)+1$		$0+z$	$x+0$			

The tame options have the same genus as those of z ($v+2$ is tame of genus 3^3 by the Intermediate Value Theorem) and v, z, x are already known to have the same species. The sum of two or any larger even number of these is therefore tame of genus 0^0, while three or any larger odd number make a tame game of genus 1^1.

The game $\mathrm{V}_{32} = y = z_1 2_2 20$ is also restive, but of a different genus, 1^3. These games $z, y, x, w, \ldots$ are still half tame, but their analysis is much more difficult than that for the first version:

	$y+z, y+x, y+v$	have genus 0^{20},
	$z+w, z+u$	have genus 3^{31},
	$y+w, y+u$	have genus 3^{13},
	$z+y+z, z+y+x$	have genus 1^{31},
while	$z+y+y$	is tame of genus 1^1.

Dawson, Officers, Grundy

For these three well known wild dogs we have made extended calculations and present you with the resulting D, O, G sequences. Shorter tables covering all the two-digit octal games will be found in the Extras.

For **Dawson's Kayles** the genus sequence for the positions D_n with n pins is:

n	1	2	3	4	5	6	7	8	9	10	11	12	13	14
D_n	0	1	1	2	0	3	1	1	0	3^{1431}	3	2^{0520}	2	4^{146}
D_{n+14}	0	5^{057}	2^{0520}	2	3^{1431}	3	0^{02}	1^{031}	1^{13}	3^{1431}	0^{31}	2^{0520}	1^{431}	1^{13}
D_{n+28}	0^{120}	4^{0564}	5^{057}	2^{20}	7^{14875}	4^{57}	0^{02}	1^{031}	1^{13}	2^{1420}	0^{31}	3^{0631}	1^{431}	1^{13}

and here is a short table for sums of two such positions:

+	A	B	C	D	D_{21}	D_{22}	D_{23}	D_{24}	D_{25}	D_{26}
$D_{10} = D_{17} + 1 = 2, 21 = A$	0^{120}	1^{031}	7^{175}	6^{064}	3^{31}	2^{0520}	2^{20}	0^{120}	3^{564}	1^{031}
$D_{12} = D_{19} + 1 = A3_20 = B$	1^{031}	0^{120}	6^{064}	7^{175}	2^{20}	3^{1431}	3^{31}	1^{031}	2^{431}	0^{120}
$D_{14} = BA_12_231 = C$	7^{175}	6^{064}	0^{120}	1^{031}	4^{5864}	5^{057}	5^{4975}	7^{175}	4^{5864}	6^{064}
$D_{16} = CB_1A_220 = D$	6^{064}	7^{175}	1^{031}	0^{120}	5^{4975}	4^{146}	4^{5864}	6^{064}	5^{20}	

The genus of any sum of terms A,B,C,D, is correctly evaluated by *pretending* that

$$A + A = B + B = C + C = D + D = 0,$$

$$A + 1 = B, \qquad C + 1 = D.$$

Dawson, whose original game *was* the misère play version, found a tendency to period 14 in the outcomes of D_n which our table verifies up to $n = 42$. But (unlike D_2, D_{16} and D_{30}) D_{44} is a misère $\mathcal{N}$-position because there is a move to $D_{21} + D_{21}$ which has genus 0^{02}, since every option has a reverting move into this genus:

$$D_{21} + D_{21} = \{D_{21} + D + 1,\ D_{21} + C, D_{21} + B + 1, D_{21} + A, D_{21} + 3 + 2, D_{21} + 2\}$$

$$D_{21} + A + 3 \qquad D_{21} + 2 + 2 \qquad 2 + 2.$$

For **Officers** (**.6**, Chapter 4) the genus sequence O_n for officers of rank n, directly responsible for $n + 2$ other officers and men, is:

n	0	1	2	3	4	5	6	7	8	9	10	11
O_n	0	0	1	2	0	1	2	3^{1431}	1	2	3^{1431}	4^{0564}
O_{n+12}	0^{20}	3^{1431}	4^{0564}	2^{20}	1^{13}	3^{0531}	2^{20}	1^{13}	0^{02}	2^{20}	1^{13}	4^{0564}
O_{n+24}	5^{475}	1^{13}	4^{0564}	5^{475}	1^{13}	2^{20}	0^{02}	1^{13}	...			

For the single heap positions G_n in **Grundy's Game** the genus sequence was computed to 50 terms in ONAG:

n	1	2	3	4	5	6	7	8	9	10
G_n	0	0	1	0	2	1	0	2	1	0
G_{n+10}	2	1	3^{1431}	2	1	3^{1431}	2	4^{0564}	3^{1431}	0^{20}
G_{n+20}	4^{0564}	3^{1431}	0^{20}	4^{0564}	3^{1431}	0^{20}	4^{0564}	1^{13}	2^{20}	3^{02}
G_{n+30}	1^{13}	2^{20}	4^{02}	1^{13}	2^{475}	4^{02}	1^{13}	2^{475}	4^{0564}	1^{13}
G_{n+40}	5^{475}	4^{0564}	1^{13}	5^{475}	4^{0520}	1^{13}	5^{475}	4^{20}	1^{13}	0^{0431}...

and Dean Alleman extended this to ranges of 96 beans and a few larger heaps.

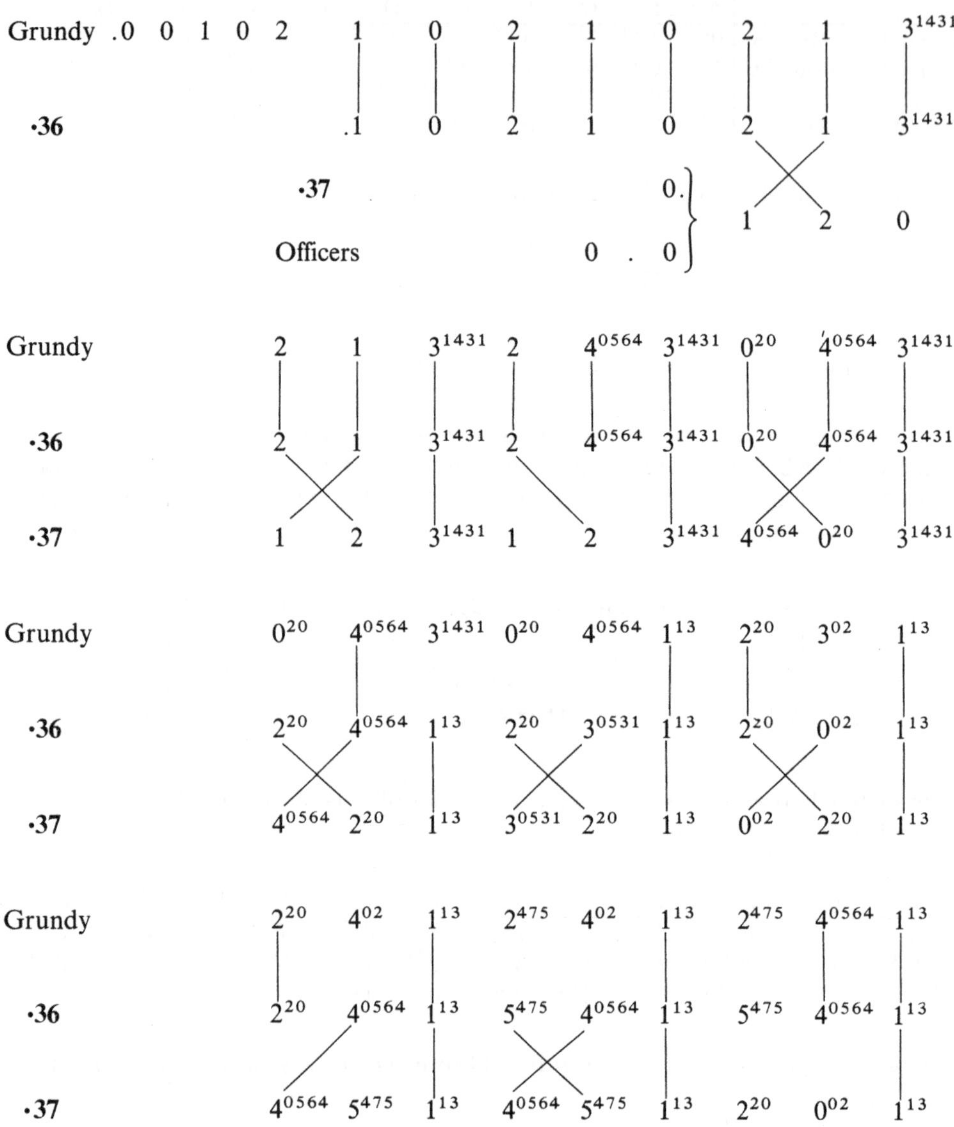

When there are no heaps of 28 or more we can *pretend* that

$$G_{13} = G_{16} = G_{19} = G_{22} = G_{25} = a,$$

$$G_{18} = G_{21} = G_{24} = G_{27} = b$$

$$G_{20} = G_{23} = G_{26} = c,$$

and that

$$a + a = b + b = 0, \qquad c + c + c = c + c.$$

When adding these, use the table:

number of copies of c:	none	one	two or more
together with:			
0	0^{120}	0^{20}	0^{02}
a	3^{1431}	3^{431}	3^{31}
b	4^{0564}	4^{564}	4^{46}
$a + b$	7^{05875}	7^{5875}	7^{75}

The following equalities show that this table is also useful when playing Officers:

$$G_{13} = G_{16} = G_{19} = a = O_7 = O_{10},$$

$$G_{18} = G_{21} = b = O_{11},$$

$$G_{20} = c = O_{12},$$

$$G_{22} = d = O_{13}.$$

In the sense of the Extras to Chapter 4, Officers is a first cousin to **·37**. The game **·36** is even more closely related to Grundy's Game, as shown in the scheme opposite.

Solid lines indicate equalities of genus which in the first two blocks are identities.

The misère theory of impartial games is the last and most complicated theory in this book. Congratulations if you've followed us this far. In the next volumes you can relax while learning to play some particular games.

Extras

All Subtraction Games Reduce to Nim

Take-A-Square, the last game in the good child's zoo, is the subtraction game corresponding to the set

$$\{1, 4, 9, 16, ...\}.$$

Although its values will never become periodic we do know that they will always reduce to Nim-heaps.

T.S. Ferguson proved from his Pairing Property that a heap of size n in the subtraction game corresponding to *any* subtraction set

$$\{s_0, s_1, s_2, ...\} \qquad (0 < s_0 < s_1 < s_2 < ...)$$

reduces to a Nim-heap of size $\mathcal{G}(n)$, its (normal) nim-value.

Certainly

$$\mathcal{G}(n) = \text{mex}\,(\mathcal{G}(n - s_0), \mathcal{G}(n - s_1), \mathcal{G}(n - s_2), ...)$$

so we need only show that if $\mathcal{G}(n) = 0$, then $\mathcal{G}(n - s_k) = 1$ for some s_k. But if $\mathcal{G}(n) = 0$ and there is *any* legal move, it is legal to take the *least* number, s_0, so $\mathcal{G}(n - s_0) > 0$. Hence there is *some* s_k with $\mathcal{G}(n - s_0 - s_k) = 0$. But we showed in the Extras to Chapter 4 that every nim-value 0 in a subtraction game is accompanied by a nim-value 1 that occurs s_0 places later, so

$$\mathcal{G}(n - s_0 - s_k + s_0) = \mathcal{G}(n - s_k) = 1,$$

as desired.

You can now use the tables in Chapter 4 to play the misère forms of all the subtraction games dealt with there.

Prim and Dim

These games from Chapter 4 are rather like subtraction games. For Prim$^+$, Prim$^-$, and Dim$^+$, every heap has an option 0 or 1 and so reduces to a Nim-heap. But for Dim$^-$ (remove from n a divisor less than n) the genus sequence is

$$.0 \quad 1 \quad 0 \quad 2 \quad 0^0 \quad ,1^3 \quad 0^0 \quad 3^{13}0^0 \quad 1^{20}0^0 \quad 1^{46}0^0 \quad 2^{13}0^0 ... \quad .$$

For $n > 5$ it can be shown that the genus is

$$g^{\gamma\delta\gamma\delta}...$$

where $n = 2^g d$ (d odd) and

$$\begin{aligned} \gamma &= g - 2 && \text{if } d = 1, \\ \gamma &= g + 2 && \text{if } d = 3, \\ \gamma &= g + 1 && \text{if } d = 5, \\ \gamma &= g && \text{if } d \geq 7, \end{aligned}$$

and

$$\delta = \gamma \overset{*}{+} 2.$$

Proof of the Noah's Ark Theorem

It remains to show that, in the conditions of the theorem, if

$$\mathrm{R} = \{\gamma, \delta^\delta, a^a, b^b, ...\}$$

is a restless game with genus $g^{\gamma\cdots}$, where γ, δ are 0,1 in some order and $2 \leq a < b < ...$, then pairs $\mathrm{R} + \mathrm{R}$ may be neglected in calculating the outcome of

$$\mathrm{R} + \mathrm{R} + \mathrm{R} + ... + \mathrm{T}$$

where T is a tame game. We prove the result by induction. If the number of copies of R is odd, we note that the options of $(2n + 1).\mathrm{R} + \mathrm{T}$,

$$2n.\mathrm{R} + \mathrm{R}' + \mathrm{T} \text{ and } (2n + 1).\mathrm{R} + \mathrm{T}'$$

have the same outcomes as

$$\mathrm{R}' + \mathrm{T} \text{ and } \mathrm{R} + \mathrm{T}',$$

the options of $\mathrm{R} + \mathrm{T}$. In particular $(2n + 1).\mathrm{R} + \gamma$ is a $\mathcal{P}$-position.

If the number of copies of R is even, it suffices to prove that $(2n+2).\mathrm{R}+\mathrm{T}$ is a $\mathcal{P}$-position when T is. If T has genus 1^0 it must be 1, and the options of $(2n + 2).\mathrm{R} + 1$ can be reverted to $\mathcal{P}$-positions:

$$(2n+2).\mathrm{R} + 1 \begin{cases} \nearrow (2n+2).\mathrm{R} \longrightarrow (2n+1).\mathrm{R} + \gamma, \\ \rightarrow (2n+1).\mathrm{R} + \gamma + 1 \longrightarrow (2n+1).\mathrm{R} + \gamma, \\ \searrow (2n+1).\mathrm{R} + \delta^\delta + 1 \longrightarrow 2n.\mathrm{R} + \gamma + \delta^\delta + 1 = 2n.\mathrm{R} + 0^0, \\ \searrow (2n+1).\mathrm{R} + a^a + 1 \searrow \left\{ \begin{array}{l} (2n+1).\mathrm{R} + \delta + 1 = (2n+1).\mathrm{R} + \gamma, \\ \text{or} \\ 2n.\mathrm{R} + a^a + (a \overset{*}{+} 1)^{a \overset{*}{+} 1} + 1 = 2n.\mathrm{R} + 0^0, \end{array} \right. \end{cases}$$

according to which part of condition (ii) of the theorem is satisfied.

The options of $(2n + 2).\mathrm{R} + \mathrm{T}$, where T has genus 0^0, can be similarly reverted:

$$(2n+2).\mathrm{R} + \mathrm{T} \begin{cases} \nearrow (2n+1).\mathrm{R} + \mathrm{R}' + \mathrm{T} \longrightarrow 2n.\mathrm{R} + \mathrm{R}' + \mathrm{R}' + \mathrm{T} = 2n.\mathrm{R} + 0^0, \\ \searrow (2n+2).\mathrm{R} + \mathrm{T}' \longrightarrow (2n+2).\mathrm{R} + \mathrm{T}'', \end{cases}$$

where T″ is a $\mathcal{P}$-position.

	0	1	2	3	4	5	6	7
·0	$.\dot{0}$	$.01\dot{0}$	$.\dot{0}01\dot{1}$	$.\dot{0}11\dot{0}$	M	$.\dot{0}\dot{1}$	M	M
·1	$.1\dot{0}$	$.11\dot{0}$	$.\dot{1}00\dot{1}$	$.\dot{1}10\dot{0}$	M	M	M	M
·2	$.\dot{0}\dot{1}$	$.\dot{0}\dot{1}$	$.\dot{0}1\dot{2}$	$.\dot{0}1\dot{2}$	$.\dot{0}\dot{1}$	$.\dot{0}\dot{1}$	M	**·26**
·3	$.\dot{1}\dot{0}$	M	$.\dot{1}0\dot{2}$	$.\dot{1}2\dot{0}$	M	M	M	M
·4	**·07**$_1$	**·17**$_1$	**·07**$_1$	**·17**$_1$	M	M	**·44**	**·45**
·5	$.\dot{1}\dot{0}$	$.\dot{1}$	M	M	M	$.\dot{1}$	M	M
·6	**·37**$_1$	**·37**$_1$	**·37**$_1$	**·37**$_1$	M	**·64**	**·64**	**·64**
·7	$.\dot{1}\dot{0}$	M	M	M	M	M	M	M
4·	$.\dot{0}\dot{1}$	$.\dot{1}$	$.\dot{0}\dot{1}$	M	**·77**$_1$	$.\dot{1}$	**·77**$_1$	M

Guide to Table 5

Misère Octal Games

Table 5 gives the genus sequence for a number of 2-digit octal games. Use the guide below for games that do not appear in the main table (M). An entry in the guide is then an equivalent game (subscript 1 denotes "first cousin of') that *does* appear, or the complete value sequence when all positions are Nim-heaps.

In Table 5, $a, b, c, ...$ refer to the first, second, third, ... entries that are not single digits. The letters A,D,T,H,F in the last column refer to the following

Notes

A Additional information is to be found in the text.
·07 = Dawson's Kayles, **·15** = Guiles, **·31** = Stalking,
·36 (cf Grundy and) **·37** = Officers, **·52** = Jelly Beans,
·56 = Lemon Drops, **·73** = Adders, **·77** = Kayles,

D Every entry is duplicated. For **·44** the duplicated values are those of Kayles.

T Every position is tame. For **·73** and **4 · 3** every position is a Nim-position. T.S. Ferguson observed that **· 73** and **·333** do *not* have the same misère play (cf. entry **·26** in Table 6(b) of Chapter 4).

H Every position is half-tame. The first few values for the three games **·26**, **·57** and **4·7** are closely related and contain increasing numbers of genus superscripts before settling down. Does this continue?

F Jim Flanigan has worked out a complete analysis:

After its first 6 values, **·34** has the exact period

$$0\,3\,1\,2_2 1\,1\,2\,0\,(2_2 1)+1$$

of eight terms. Two copies of $2_2 1$ can be neglected by the Noah's Ark Theorem.

After its first 2 terms, the genus sequence for **·71** exhibits period 6:

$$1\,0\,1^{20}\,0\,1\,0^{1420}.$$

Although the various games of genus 1^{20} and 0^{1420} are distinct we obtain a correct analysis by *pretending* they are equivalent to r and s and using the genus table:

$+$	0	r	$2r$	$3r, 5r, 7r, \ldots$	$4r, 6r, 8r, \ldots$
$0, 2s, 4s, \ldots$	0^{120}	1^{20}	0^{02}	1^{13}	0^{02}
$s, 3s, 5s, \ldots$	0^{1420}	1^{02}	0^{20}	1^{13}	0^{02}

Game	Genus-sequence		Notes
·04	$.00011\ 12203\ 31110\ 4^{146}3^{1431}332^{0520}\ 224^{146}4^{146}0^{120}$	$a=2_2321\ \ b=2_221$	
·06	$.00112\ 20311\ 22^{1420}3^{1431}34^{0564}\ 4^{0564}0^{20}5^{1475}3^{1431}$	$a=2_231\ \ b=2_221$	
·07	$.01120\ 31103^{1431}\ 32^{0520}24^{146}0\ 5^{057}2^{0520}23^{1431}3\ 0^{02}$	$a=2_221\ \ b=a3_20$	A
·14	$.10010\ 21221\ 0414^{146}4\ 1^{3}2^{1420}21^{3}2^{1720}\ 0^{02}1^{031}0^{120}$	$a=2_2321\ \ b=2_2420$	
·15	$.11011\ 22122\ 1101^{4}1^{4}\ 221^{2}22^{1420}\ 1^{4}1^{4}0^{1}1^{4}1^{4}$	$a=b=2_2320\ \ c=a_12_20$	A
·16	$.10012\ 21401\ 42^{1420}1^{2}4^{2046}0^{1}\ 1^{0}4^{031}2^{1657}2^{5720}4^{20}$	$a=2_2431\ \ b=2_240$	
·17	$.11021\ 30113^{1431}\ 22^{0520}34^{146}1\ 5^{057}3^{1431}22^{0520}3\ 1^{13}$	$a=2_221\ \ b=a3_20$	
·26	$.01230\ 1230^{4}1^{5}\ 2^{0420}3^{31}0^{20420}1^{531}2^{020420}\ 3^{1431}$	$a=3_2321\ \ b=a+1\ \ c=a_1a3$	H
·31	$.120^{0}1^{1}0^{0}\ 1^{1}0^{0}1^{1}0^{0}1^{1}\ 0^{0}1^{1}0^{0}1^{1}0^{0}$	$a=2_+\ \ b=2_{++}$	AT
·34	$.10120\ 10312^{1420}\ 1203^{0531}0\ 312^{1420}12\ 03^{0531}$	$a=c=2_21\ \ b=d=a+1$	F
·35	$.120^{0}1^{2}0^{0}\ 2^{2}1^{02}2^{2}0^{0}1^{2}\ 0^{0}2^{2}1^{02}2^{2}0^{0}\ 1^{20}0^{0}2^{2}$	$a=2_+\ \ b=2_+0$	H
·36	$.10210\ 213^{1431}21\ 3^{1431}24^{0564}3^{1431}0^{20}\ 4^{0564}3^{1431}2^{20}$	$a=b=d=2_221\ \ c=f=a_2a20$	A
·37	$.12012\ 3^{1431}123^{1431}4^{0564}\ 0^{20}3^{1431}4^{0564}2^{20}1^{13}\ 3^{0531}$	$a=b=2_221\ \ c=a_2a20$	A
·44	$.00112\ 23311\ 4^{146}4^{146}332^{2}\ 2^{2}1^{1}1^{1}4^{046}4^{046}$	$a=b=2_2321\ \ c=d=a_1a3_22_230$	D
·45	$.01122\ 3114^{146}4^{4}\ 32^{2}2^{2}1^{1}1^{0}\ 4^{057}2^{2}$	$a=2_2321\ \ b=2_23210$	
·52	$.10221\ 0^{0}32^{2}1^{1}0^{0}\ 3^{3}2^{2}1^{1}0^{0}3^{3}\ 2^{2}1^{1}0^{0}$	$a=2_2\ \ b=3_22_21\ \ c=3_2\ \ d=b_1b2_{22}3$	AT
·53	$.11221\ 0^{0}22^{2}4^{4}0^{0}\ 1^{6}2^{2}2^{2}1^{031}1^{1}\ 2^{2}2^{2}4^{4}1^{1}1^{631}$	$a=2_2\ \ b=3_22_20\ \ c=3_22_232$	H
·54	$.10122\ 2411^{1}1^{6}\ 222^{2}4^{157}1^{1}\ 1^{1}1^{431}2^{2}2^{2}$	$a=2_2532\ \ b=2_254320$	H
·56	$.10224\ 11^{1}324^{4}\ 4^{4}6^{6}6^{6}2^{2}1^{1}\ 1^{1}7^{7}6^{6}8^{8}4^{4}\ 1^{1}1^{1}$	$a=2_25432\ \ b=a_1a4_22_2321$	AT
·57	$.11221\ 1221^{4}1^{4}\ 2^{20}2^{20}1^{431}1^{431}2^{1420}\ 2^{1420}1^{20431}$	$a=b=2_2320\ \ c=d=a_1a2_230$	DH
·64	$.01234\ 15^{146}3^{3}21^{431}\ 5^{146}4^{657}2^{7846}6^{4657}$	$a=2_24321\ \ b=a3_22_25421$	
·71	$.12101^{2}\ 010^{1420}10\ 1^{20}010^{1420}1\ 01^{20}010^{1420}$	$a=2_20\ \ b=a_21$	F
·72	$.10231\ 0231^{4}0^{0}\ 2^{20}3^{31}1^{431}0^{0}2^{1420}$	$a=2_2320\ \ b=2_2\ \ c=a_2a2_230$	H
·73	$.1230^{0}1^{1}\ 2^{2}3^{3}0^{0}1^{1}2^{2}\ 3^{3}0^{0}1^{1}2^{2}3^{3}\ 0^{0}1^{1}2^{2}$	$a=2_2\ \ b=3_2\ \ c=2_{22}\ \ d=3_{22}$	AT
·74	$.10123\ 2414^{146}6^{657}\ 232^{2}1^{1}5^{857}\ 1^{1}7^{657}$	$a=2_25321\ \ b=a3_22_254320$	
·75	$.12121^{4}\ 2^{20}1^{431}2^{20}1^{1}2^{20}\ 1^{13}2^{20}1^{13}2^{20}$	$a=2_2320\ \ b=a2_230$	H
·76	$.10234\ 16^{146}234^{46}\ 1^{531}6^{146}7^{657}3^{3}2^{20}\ 1^{531}6^{746}$	$a=2_254321\ \ b=a_1a4_34_22_2321$	
·77	$.12314^{146}\ 32^{2}1^{1}4^{046}2^{2}\ 6^{46}4^{046}1^{1}2^{2}7^{57}\ 1^{13}4^{64}3^{31}2^{20}$	$a=2_2321\ \ b=a_1a3_22_230$	A
4·3	$.120^{0}2^{2}0^{0}\ 2^{2}0^{0}2^{2}0^{0}2^{2}\ 0^{0}2^{2}0^{0}2^{2}0^{0}\ 2^{2}$	$a=2_2\ \ b=2_{22}\ \ c=2_{222}$	T
4·7	$.12121^{4}\ 2^{20}1^{431}2^{1420}1^{20431}2^{131420}\ 1^{2020431}2^{13131420}$	$a=2_2320\ \ b=a_1a2_230$	H

Table 5. Genus Sequences of Two-Digit Octal Games.

Stop Press: Even More Games are Tameable!

A *single* tameable game can be added to any number of tame ones by the usual rules. If

$$G = \{T_1, T_2, ..., U_1, U_2, ...\}$$

where $T_1, T_2, ...$ are games we already know to be tameable, while $U_1, U_2, ...$ may be wild, then the three conditions for G to be **tameable** are:

(i) its nim-values g^γ must be one of the tame pairs

$$0^0, 1^1, 2^2, 3^3, 4^4, ... \text{ or } 0^1 \text{ or } 1^0$$

(ii) g^γ must also be the nim-values of the **reduced game**

$$H = \{T_1, T_2, ...\}$$

(except that when H is empty, g^γ may be 0^1 *or* 0^0)

(iii) from each of the wild or unknown options $U_1, U_2, ...$ there must be reverting moves to two (possibly equal) tameable games of genus

$$g^? \quad \text{and} \quad ?^\gamma$$

For example, all the entries

$$0^{02}, 1^{13}, 2^{30}, 3^{31}, 0^{120}, 1^{031}$$

for the Kayles positions in Table 4 are tameable, of genus

$$0^{0/}, 1^{1/}, 2^{2/}, 3^{3/}, 0^{1/}, 1^{0/}$$

(/ being our way of distinguishing TAMEABLE from TAME). We have TAMEABLE + TAME = TAMEABLE, but TAMEABLE + TAMEABLE may be WILD. In fact TAME is the largest class of games that can be added like Nim-positions, and in which one player has a winning strategy which always returns to the class.

A Complete Strategy for Misère Kayles

This remarkable strategy has been found by William Sibert since our first edition.

Let us first remind ourselves how to play Kayles well under the normal play rules. Figure 8 should enable you to find the normal play outcome of any Kayles position without performing any nim-additions (or even understanding what they mean). Each of the circled nodes 1, 2, 3, 4, 5, 6, 7, 8, represents a nim-value, and near it are written the lengths of all Kayles-rows having that nim-value. A superscript + sign indicates that arbitrary multiples of 12 may be added.

We'll say that a Kayles position **mentions** a node as many times as it has rows of the lengths written near that node. Then, to win in normal play, you should move only to positions of nim-value **zero**, that is to positions that

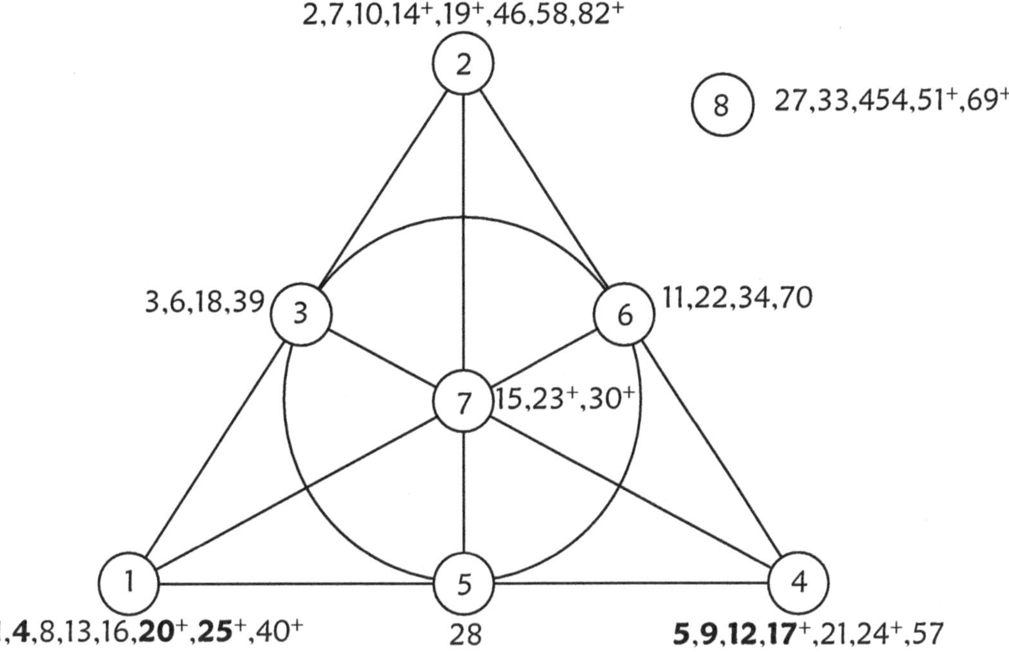

Figure 8. How to win in normal play Kayles.
(n^+ means n, $n+12$, $n+24$, $n+36$, ...)

(A) *mention no node oddly* (i.e. an odd number of times), *or*
(B) *mention oddly just three nodes in line* (note that 3, 5, 6 are in line), *or*
(C) *mention oddly just four nodes, namely, all except three in line and* 8, *or*
(D) *mention oddly all seven nodes*, 1, 2, ... , 7.

Following a time-honored practice we imagine that players can analyze one move ahead and so regard a game as solved when we have given a rule to determine the outcome of any given position. In fact, finding the good moves from an $\mathcal{N}$-position can be hard, but in normal play you can restrict the search to those options which involve a component of maximal nim-value.

To understand our misère strategy, first note that each Kayles position is of one of four types:

NN: $\mathcal{N}$-positions in both normal and misère play.
PP: $\mathcal{P}$-positions in both normal and misère play.
PN: $\mathcal{P}$-positions in normal play, but $\mathcal{N}$-positions in misère play.
NP: $\mathcal{N}$-positions in normal play, but $\mathcal{P}$-positions in misère play.

Positions of types NN and PP which behave similarly in normal and misère play are called **frigid**; those of types PN and NP which behave differently in normal and misère play are called **frisky**. Our claim is that a position is frisky only if all of its non-empty rows are of

To win at misère Kayles, move as you would in normal Kayles, **except** that **you must not** move to a position of type PN:
The **only** positions of type PN are of one of the three forms

(α) E(5)·E(4,1),
(β) E(17,12,9)·E(20,4,1), or
(γ) 25·E(17,12,9)·D(20,4,1).

All positions of form (α), (β) or (γ) are of type PN and each of them has nim-value 0. Note that the empty position is included in (α) and (β).
On the other hand **you may** move to a position of type NP:
The positions of type NP are **exactly those** of the six forms

D(5)·D(4,1)	with nim-value	5
E(5)·D(4,1)	...	1
D(9)·E(4,1)	...	4
12·E(4,1)	...	4
E(17,12,9)·D(20,4,1)	...	1
25·D(9)·D(4,1)	...	4

Table 6. A Strategy for Misère Kayles.

risky length, i.e., of length 1, 4, 5, 9, 12, 17, 20 or 25 (the **bold** numbers in Fig. 8). We also claim that all frisky positions are of one of the forms in Table 6. In this the notation D$(a, b, \ldots)$ resp. E$(a, b, \ldots)$ refers to collections of oddly resp. evenly many rows of lengths a or b or $\ldots$. For example, $5 \cdot 4 \cdot 4 \cdot 1$ is of form D(5)·E(4) while $4 \cdot 1 \cdot 1$ is of form D(4)·E(1).

It is also helpful to list some frigid positions:

type	notation	description
NN	(n)	any position of non-zero nim-value containing a row of **rigid** (non-risky) length $n \neq 1, 4, 5, 9, 12, 17, 20, 25$.
NN	(D·E)	a position of form D(5)·E(4,1), nim-value 4.
NN	(D·D)	a position of form D(17,12,9)·D(25,20,4,1), nim-value 5.
NN or PP	$(5 \cdot N)$	any position containing a row of 5, and a longer row.
PP	$(25 \cdot 25)$	a position with just two rows of 25, nim-value 0.

Our proof of the foregoing statements is via a set of four tables:

Table PN. Table PN verifies the strategy for a non-empty position asserted to be of type PN. For the next player to win, it suffices to find a move to a position of type NP. These are listed in Table PN.

Form of PN position	Recommended move ...	yielding an NP position of form
(α) E(5)·E(4,1)	$4 \to 1 \cdot 1$ or $1 \to 0 \cdot 0$	E(5) · D(4,1)
	else $5 \to 4 \cdot 0$	D(5) · D(4,1)
(β) E(17,12,9)·E(20,4,1)	$\{ 20 \to 9 \cdot 9, 17 \to 12 \cdot 4, 12 \to 9 \cdot 1$; $4 \to 1 \cdot 1$ or $1 \to 0 \cdot 0 \}$	E(17,12,9)·D(20,4,1)
	else $9 \to 4 \cdot 4$	D(9)·E(4,1)
(γ) 25·E(17,12,9)·D(20,4,1)	$25 \to 12 \cdot 12$	E(17,12,9)·D(20,4,1)

Table PN. Recommended Moves from a Non-Empty Position of type PN.

Table NP. For a position asserted to be of type NP we must check that there is no move to a position of type PP or NP. Since all positions of type PP have nim-value 0, and those of type NP have nim-value 1, 4 or 5, the change, Δ, in the nim-value for such a move must be 1, 4 or 5. Table NP has a row for each move $n \to a \cdot b$, for which n is a risky number, 1, 4, 5, 9, 12, 17, 20 or 25 and $\Delta = 1, 4$ or 5, and a column for each of the six forms of NP position. The entries are either N.A., if the move is not applicable; or (α), (β) or (γ) if the move is to a PN position of that form; or (n), ($5 \cdot N$), ($25 \cdot 25$), (D·E) or (D·D) if the move is to a frigid position of that form (not of type PP since the nim-value is not zero).

Move	with $\Delta =$	from a position of type NP, and of form (and nim-value $\mathcal{G}$)					
		E(17,12,9) ·D(20,4,1)	E(5)· D(4,1)	25·D(9) ·D(4,1)	12· E(4,1)	D(9)· E(4,1)	D(5)· D(4,1)
		$\mathcal{G} = 1$	$\mathcal{G} = 1$	$\mathcal{G} = 4$	$\mathcal{G} = 4$	$\mathcal{G} = 4$	$\mathcal{G} = 5$
$1 \to 0 \cdot 0$ or $4 \to 1 \cdot 1$	1	(β)	(α)	(D·D)	(D·D)	(D·D)	(D·E)
$12 \to 9 \cdot 1$	1	(β)	N.A.	N.A.	(D·D)	N.A.	N.A.
$17 \to 12 \cdot 4$ or $20 \to 9 \cdot 9$	1	(β)	N.A.	N.A.	N.A.	N.A.	N.A.
$25 \to 20 \cdot 4$ or $12 \cdot 12$	1	N.A.	N.A.	(D·D)	N.A.	N.A.	N.A.
$25 \to 18 \cdot 6$, $16 \cdot 8$ or $14 \cdot 10$	1	N.A.	N.A.	(n)	N.A.	N.A.	N.A.
$5 \to 2 \cdot 2$	4	N.A.	(2)	N.A.	N.A.	N.A.	(2)
$9 \to 4 \cdot 4$	4	(D·D)	N.A.	(γ)	N.A.	(β)	N.A.
$12 \to 5 \cdot 5$	4	($5 \cdot N$)	N.A.	N.A.	(α)	N.A.	N.A.
$20 \to 17 \cdot 1$	4	(D·D)	N.A.	N.A.	N.A.	N.A.	N.A.
$17 \to 14 \cdot 2$ or $8 \cdot 8$, $20 \to 13 \cdot 5$	4	(n)	N.A.	N.A.	N.A.	N.A.	N.A.
$5 \to 4 \cdot 0$	5	N.A.	(D·E)	N.A.	N.A.	N.A.	(α)
$9 \to 8 \cdot 0$ or $6 \cdot 2$	5	(n)	N.A.	(n)	N.A.	(n)	N.A.
$12 \to 7 \cdot 3$	5	(3)	N.A.	N.A.	(3)	N.A.	N.A.
$17 \to 16 \cdot 0$ or $10 \cdot 6$, $20 \to 15 \cdot 3$ or $11 \cdot 7$	5	(n)	N.A.	N.A.	N.A.	N.A.	N.A.
$25 \to 24 \cdot 0$ or $22 \cdot 2$	5	N.A.	N.A.	(n)	N.A.	N.A.	N.A.

Table NP. There is No Move, Involving a Risky Number,from a Position of Type NP to One of Type PP or NP.

Moves	with $\Delta =$	to a position of form	must have been from a position of form
$1 \to 0 \cdot 0$ or $4 \to 1 \cdot 1$	1	E(17,12,9)·D(20,4,1) or E(5)·D(4,1)	(β) or (α)
$12 \to 9 \cdot 1$, $17 \to 12 \cdot 4$, $20 \to 9 \cdot 9$			(β) or N.A.
$25 \to 12 \cdot 12$ or $20 \cdot 4$			(γ) or N.A.
$16 \to 9 \cdot 5$, $36 \to 25 \cdot 9$, $52 \to 25 \cdot 25$			N.A.
$9 \to 4 \cdot 4$	4	12·E(4,1),	(β) or (γ)
$8 \to 5 \cdot 1, 12 \to 5 \cdot 5, 20 \to 17 \cdot 1,$ $24 \to 17 \cdot 5, 32 \to 25 \cdot 5, 36 \to 17 \cdot 17,$ $41 \to 20 \cdot 20, 44 \to 25 \cdot 17$	4	D(9)·E(4,1) or 25·D(9)·D(4,1)	N.A.
$5 \to 4 \cdot 0$	5	D(5)·D(4,1)	(α)
$13 \to 12 \cdot 0$, $21 \to 20 \cdot 0$, $28 \to 17 \cdot 9$ or $25 \cdot 1$			N.A.

Table PP. There Is No Move $n \to a \cdot b$, from a PP Position to an NP Position, with a, b Risky and $\Delta = 1$, 4 or 5.

Table PP. Here we need only check that there is no move to a position of type NP. Such a move must be of the form $n \to a \cdot b$, where a and b are risky (chosen from 0, 1, 4, 5, 9, 12, 17, 20 and 25) and Δ, the change in nim-value, is 1, 4 or 5, since all positions of types PP and NP have nim-value 0, 1, 4 or 5. All such moves appear in Table PP, and all applicable ones lead to positions of type PN, i.e. of form (α), (β) or (γ).

$a = 0$	1	4	5	9	12	17	20	25	b
NN2	NN2	NN2 $3 \cdot 2$	$5 \to 3 \cdot 0$ $7 \to 5 \cdot 1$	NN2	NN2 $11 \cdot 2$	NN2 $15 \cdot 3$	$18 \cdot 2$	$19 \cdot 6$	0
	NN2	$2 \cdot 2$ $3 \cdot 3$	$7 \to 5 \cdot 0$ $5 \to 2 \cdot 2$	NN2	$8 \cdot 5$	$13 \cdot 5$	$19 \cdot 2$	$21 \cdot 5$	1
		NN2 $8 \cdot 1$	$5 \to 3 \cdot 0$ $11 \to 9 \cdot 1$	$8 \cdot 5$	NN2 $15 \cdot 2$	$16 \cdot 5$	$18 \cdot 6$	$24 \cdot 5$	4
			$8 \cdot 1$ $5 \to 2 \cdot 2$	$(5 \cdot N)$	$(5 \cdot N)$	$(5 \cdot N)$	$(5 \cdot N)$	$(5 \cdot N)$	5
				$10 \cdot 7$ NN2	$13 \cdot 8$	$13 \cdot 13$	$18 \cdot 11$	$21 \cdot 13$	9
					$14 \cdot 10$	$16 \cdot 13$	$24 \cdot 8$	$24 \cdot 13$	12
						$29 \cdot 5$	$21 \cdot 16$	$23 \cdot 19$	17
							$32 \cdot 8$	$24 \cdot 21$	20
								$(25 \cdot 25)$	25

Table NN1.
An entry $c \cdot d$ refers to a substitute move $n \to c \cdot d$ for the risky move $n \to a \cdot b$. A single entry has $c + d = a + b$; an entry in the upper half of a cell has $c + d = a + b - 1$ and is available just if the original move knocked down one pin; an entry in the lower half of a cell has $c + d = a + b + 1$ and is available just if the original move knocked down two pins. An entry NN2 means that substitute moves are found in Table NN2, or that the position is not frisky. An entry $m \to c \cdot d$ indicates another form of substitute move which will be applicable provided the move was not from an NP position. An entry $(5 \cdot N)$ or $(25 \cdot 25)$ indicates that the move was in fact to a frigid position.

Table NN. For positions asserted to be of type NN, the normal play move, $n \to a \cdot b$, usually also works in misère play. When can it fail? Only when it is to a position of type PN, so that a and b are both risky. Table NN1 lists a substitute move for most of these cases, or it has an entry $(5 \cdot N)$ or $(25 \cdot 25)$ indicating that the position moved to is frigid. Substitute moves for the cases omitted from Table NN1 are found in Table NN2, or the position is not frisky after all.

In place of the risky move	from G (of type NN) to positions of type PN and of form		
	(α) E(5)·E(4,1)	(β) E(17,12,9)·E(20,4,1)	(γ) 25·E(17,12,9)·D(20,4,1)
$1 \to 0 \cdot 0$	G was NP	G was NP	$25 \to 14 \cdot 10$
$2 \to 0 \cdot 0$	$2 \to 1 \cdot 0$	$2 \to 1 \cdot 0$	$25 \to 15 \cdot 9$
$2 \to 1 \cdot 0$	$2 \to 0 \cdot 0$	$2 \to 0 \cdot 0$	$25 \to 20 \cdot 3$
$3 \to 1 \cdot 0$	$3 \to 1 \cdot 1$	$3 \to 1 \cdot 1$	$25 \to 15 \cdot 9$
$3 \to 1 \cdot 1$	$3 \to 1 \cdot 0$	$3 \to 1 \cdot 0$	$25 \to 20 \cdot 3$
$4 \to 1 \cdot 1$	G was NP	G was NP	$25 \to 14 \cdot 10$
$5 \to 4 \cdot 0$	G was NP	$9 \to 8 \cdot 0$, $12 \to 7 \cdot 3$, $17 \to 16 \cdot 0$, $20 \to 15 \cdot 3$ or G was NP	$25 \to 24 \cdot 0$
$9 \to 4 \cdot 4$	$5 \to 2 \cdot 2$ or G was NP	$12 \to 5 \cdot 5$, $17 \to 8 \cdot 8$, $20 \to 13 \cdot 5$ or G was NP	as for (β)
$10 \to 9 \cdot 0$	N.A.	$9 \to 7 \cdot 0$, $12 \to 10 \cdot 0$, $17 \to 13 \cdot 3$	$25 \to 21 \cdot 3$
$11 \to 9 \cdot 0$	N.A.	$9 \to 5 \cdot 2$, $12 \to 11 \cdot 0$, $17 \to 15 \cdot 1$	$25 \to 15 \cdot 9$
$11 \to 9 \cdot 1$	N.A.	$9 \to 5 \cdot 3$, $12 \to 6 \cdot 5$, $17 \to 15 \cdot 0$	$25 \to 20 \cdot 3$
$12 \to 9 \cdot 1$	N.A.	G was NP	$25 \to 14 \cdot 10$
$13 \to 12 \cdot 0$	N.A.	$9 \to 8 \cdot 0$, $12 \to 7 \cdot 3$, $17 \to 16 \cdot 0$	$25 \to 24 \cdot 0$
$17 \to 12 \cdot 4$	N.A.	G was NP	$25 \to 14 \cdot 10$
$18 \to 17 \cdot 0$	N.A.	$9 \to 7 \cdot 1$, $12 \to 10 \cdot 1$, $17 \to 14 \cdot 1$	$25 \to 23 \cdot 1$
$20 \to 9 \cdot 9$	N.A.	G was NP	$25 \to 14 \cdot 10$

Table NN2. Substitutes for risky moves from G (of type NN) to positions of type PN.

Sibert's remarkable *tour de force* raises once again the question: are misère analyses really so difficult? A referee of a draft of the Sibert-Conway paper wrote "the actual solution will have no bearing on other problems", while another wrote "the ideas are likely to be applicable to some other games". The present success seems to be due to the facts that the period of the normal analysis is not very large, that the number and lengths of the risky rows are not large, and, perhaps most importantly, that there there are no non-empty single-row $\mathcal{P}$-positions. We thought that there may not be many candidates awaiting a similar treatment, but there are the octal games **·375** and **·772** as well as **·144**, **·145**, **·147**, **·154** and their generalizations (see Table 8 of Chapter 4 and Table 5 of this chapter). In the meantime, Thane Plambeck introduced a weight function and obtained a Sibert-Conway type analysis of **4·7** (& **4·71**), of **4·74** (& **4·75**, **·75**, **·751**, **·752**, **·753**), of **·754** (& **·755**), of **·756** (& **·757**), of **·355** and of **·357**. Moreover, he conjectures that **·15**, **·26**, **·53**, **·54**, **·57**, **·72** and their cousins (Table 6(b) of Chapter 4) will yield to a similar analysis.

References and Further Reading

Dean Allemang, *Machine computation with finite games*, MSc thesis, Cambridge University, 1984.

Charles L. Bouton, Nim, a game with a complete mathematical theory, *Ann. of Math., Princeton*(2), **3**(1901-02) 35–39.

J. H. Conway, *On Numbers and Games*, A K Peters,Ltd., Natick MA, 2001.

T. S. Ferguson, on sums of graph games with the last player losing, *Internat. J. Game Theory*, **3**(1974) 159–167; *MR* **52** #5046.

P. M. Grundy & C. A. B. Smith, Disjunctive games with the last player losing, *Proc. Cambridge Philos. Soc.*, **52**(1956) 527–533; *MR* **18**, 546.

T. H. O'Beirne, *Puzzles and Paradoxes*, Oxford Univ. Press, London, 1965, pp. 131–150.

Thane E. Plambeck, Daisies, Kayles and the Sibert-Conway decomposition in misère octal games, *Theoret. Comput. Sci*, **96**(1992) 361-388.

William L. Sibert & John H. Conway, Mathematical Kayles, *Internat. J. Game Theory*, **20**(1992) 237–246.

Yōhei Yamasaki, On misère Nim-type games, *J. Math. Soc. Japan*, **32**(1980) 461–475.

Yōhei Yamasaki, The projectivity of Y-games, *Publ. RIMS*, Kyoto Univ., **17**(1981) 245–248.

Glossary

$\mathbf{A} = \textbf{ace} = \{0 \mid \textbf{tiny}\}$, 357
$\bar{\mathbf{A}} = -\textbf{ace} = \{\textbf{miny} \mid 0\}$, 359
$\mathbf{A}\text{-} = \{\textbf{on} \mid \mathbf{A} \,||\, 0\}$, 359
$\mathbf{A}+ = \{0 \,||\, \bar{\mathbf{A}} \mid \textbf{off}\}$, 359
$\aleph_0$, aleph-zero, 329
Δ, also, slow join, 286
$\wedge$, and, join, 278
$(x)_n, \{x| - y\}_n$, Childish Hackenbush values, 238
$a < b >$, class a and variety b, 363
$\bar{1}\clubsuit = \{\clubsuit \mid 0\}$, clubs, 359
$\clubsuit = 0\clubsuit = \{1\clubsuit \mid 0\}$, clubs, 359
$1\clubsuit = \{\textbf{deuce} \mid 0\}$, clubs, 359
• ● ○ ⊗, Col and Snort positions, 47, 146
γ°, degree of loopiness, 361
$2\clubsuit = \{0|\textbf{ace}\} = \textbf{ace} + \textbf{ace} = \textbf{deuce}$, 357
$\bar{1}\diamondsuit = \{\bar{\mathbf{J}}|0\}$, diamonds, 359
$\diamondsuit = 0\diamondsuit = \{\textbf{ace} \mid \bar{1}\diamondsuit\}$, diamonds, 359
$1\diamondsuit = \{0|\diamondsuit\}$, diamonds, 359
$\Downarrow = \{\downarrow *|0\} = \downarrow + \downarrow$, double-down, 68, 69, 71
$\Uparrow = \{0|\uparrow *\} = \uparrow + \uparrow$, double-up, 68, 69, 71
$\Uparrow * = \{0|\uparrow\} = \uparrow + \uparrow + *$, double-up star, 71
$\downarrow = \{*|0\}$, down, 64
$\downarrow_2 = \{\uparrow *|0\}$, down-second, 235, 236
$\downarrow_3 = \{\uparrow + \uparrow^2 + * \,|0\}$, down-third, 235, 236
$\downarrow_{abc} \ldots, \downarrow_{abc} \ldots$, 261
$+$, downsum, 336, 357
dud = {**dud**|**dud**}, **d**eathless **u**niversal **d**raw, 337
ϵ, epsilon, small positive number , 328
$\div$, equally uppity, 242, 247, 249
$\lfloor\ \rfloor$, "floor," greatest integer not greater than, 51, 311
$||0$, fuzzy, 29, 32, 37
G, general game, 28–31
$G \,||\, 0, G$ fuzzy 2nd player wins, 29
$G < 0, G$ negative, R wins, 29
$G > 0, G$ positive, L wins, 29
$G = 0, G$ zero, 1st player wins, 29
$G + H$, sum of games, 31
G^L, (set of) L option(s), 31
G^R, (set of) R option(s), 31
$\mathcal{G}(n)$ nim-value, 82
$G.\uparrow = \{G^{\mathrm{L}}.\uparrow + \Uparrow * \mid G^{\mathrm{R}}.\uparrow + \Downarrow *\}$, 247, 255, 258
$>$, greater then, 32
$\geq$ greater than or equal, 32
$\lhd|$, greater than or incomparable, 32, 35
$\dot{\geq}$, at least as uppity, 242, 246
$\frac{1}{2} = \{0|1\}$, half, 7, 20
$\bar{1}\heartsuit = \{\heartsuit|0\}$, hearts, 359
$\heartsuit = 0\heartsuit = \{1\heartsuit|\bar{\mathbf{A}}\}$, hearts, 359
$1\heartsuit = \{0 \mid \textbf{joker}\}$, hearts, 359
hi = {**on** || 0|**off**}, 355
hot = {**on**|**off**}, 355
$||$, incomparable, 35
$\infty = \mathbb{Z}||\mathbb{Z}|\mathbb{Z}$, 329, 334, 391
$\pm\infty = \infty| - \infty = \mathbb{Z}|\mathbb{Z} = \int^{\mathbb{Z}} *$, 334
$\infty \pm \infty = \infty|0 = \mathbb{Z}|0$, 334
$\infty + \infty = 2.\infty = \mathbb{Z} \,||\, \mathbb{Z}|0$, double infinity, 334
$\infty_0 \cdot \bar{\infty}_2$, unrestricted tallies, 314
$\infty_{abc\ldots}$, 387–395
$\infty_{\beta\gamma\delta\ldots}$, 333
$\int$, integral, 167–180
$\mathbf{J} = \{0|\bar{\mathbf{A}}+\} = \textbf{ace} + (-\textbf{ace}) = \textbf{joker}$, 358
$\bar{\mathbf{J}} = \{\bar{\mathbf{A}} - |0\} = \textbf{ace} + (-\textbf{ace}) = -\textbf{joker}$, 359
L, Left, 2
L*n*L, L*n*R, R*n*R, positions in Seating games, 44
$<$, less than, 32
$\leq$ less than or equal, 32

$\lhd\!|$, less than or incomparable, 32, 35
$\mathbf{lo} = \{\mathbf{on}|0 \,||\, \mathbf{off}\}$, 355
☽, loony, 397, 407
$\gamma, \gamma^*, \gamma^+, \gamma^-$, loopy games, 335
$s\&t$ loopy games, 336
$-1 = \{\ |0\}$, minus one, 19
$-_{\mathbf{on}} = \{\mathbf{on}|0 \,||\, 0\}$, **miny**, 353
$-_{\frac{1}{2}} = \{\frac{1}{4}|0 \,||\, 0\}$, miny-a-quarter, 134
$-_x = \{x|0 \,||\, 0\}$, miny-x, 126
$\overset{*}{+}$ nim-sum, 58
$\mathbf{off} = \{\ \ |\mathbf{off}\}$, 336–340, 357
$\omega = \{0, 1, 2, \ldots|\ \}$, 329–333
$\omega + 1 = \{\omega|\ \ \}$, omega plus one, 329–333
$\omega\times 2 = \{\omega, \omega+1, \omega+2, \ldots|\ \} = \omega+\omega$, 329–333
$\omega^2 = \{\omega, \omega\times 2, \omega\times 3, \ldots|\ \} = \omega\times\omega$, 329–333
$\mathbf{on} = \{\mathbf{on}|\ \ \}$, 336–341
$1 = \{0|\ \ \}$, one, 7, 19
$\mathbf{ono} = \{\mathbf{on}|0\}$, 355–356
$\mathbf{oof} = \{0|\mathbf{off}\}$, 355-357
$\mathbf{over} = \{0|\mathbf{over}\} = \frac{1}{\mathbf{on}}$, 341
$\pi = 3.141592653\ldots$, pi, 328–329
$\pm 1 = \{1|-1\}$, plus-or-minus-one, 120–122
$(\pm 1).\uparrow - \{*|*\}$, 247
$\Downarrow\!\!\Downarrow = \{\Downarrow *|0\} = 4.\downarrow$, quadruple-down, 347
$\Uparrow\!\!\Uparrow = \{0|\Uparrow *\} = 4.\uparrow$, 71, 347
$\frac{1}{4} = \{0|\frac{1}{2}\}$, quarter, 6, 20
$\frac{1}{4} = \frac{1}{4}.\uparrow = \{\Uparrow *|1\frac{1}{2}.\downarrow + *\}$, quarter-up, 236
$\frac{1}{4}* = \frac{1}{4}.\uparrow + * = \{\Uparrow |1\frac{1}{2}.\downarrow\}$, quarter-up-star, 236
R, Right, 2
$\frac{*}{2} = \{*, \uparrow | \downarrow *, 0\}$, semi-star, 370
$\frac{1}{2} = \frac{1}{2}.\uparrow = \{\Uparrow *|\downarrow *\}$, semi-up, 236, 247
$\frac{1}{2}* = \frac{1}{2}.\uparrow + * = \{\Uparrow | \downarrow\}$, semi-up-star, 236
$\frac{3}{2} = 1\frac{1}{2}.\uparrow = \{*|*\}$, sesqui-up, 236
sign(), 348–350
$|, ||, |||, \ldots$ slash, slashes, ... separate L and R options, 6–7, 128–129, 366
$2♠ = \{\bar{\mathbf{A}}|0\}$, spades, 359
$\bar{1}♠ = \{0|\bar{2}♠\}$, spades, 359
$♠ = 0♠ = \{0|\bar{1}♠\}$, spades, 359
$1♠ = 0|♠$, spades, 359
☆, far star, remote star, 230–232, 244–251
$* = \{0|0\}$, star, 38
$*2 = \{0, *|0, *\}$, star-two, 41
$*n = \{0, *, \ldots, *(n-1)|0, *, \ldots, *(n-1)\}$, star-$n$, 41
$*\alpha$, star-alpha, 333
$\hat{*} = *.\uparrow = \{\Uparrow *| \downarrow *\}$, starfold-up, 236, 248
$*\bar{n}$, all nimbers except $*n$, 397
$*n \rightarrow .$, all nimbers from $*n$ onwards, 397
$\odot = 0* \rightarrow$, sunny, 397–401, 404
$\uparrow_{abc\ldots} = -\downarrow^{abc\ldots}$, superstars, 261
$2_1 3_0$, tallies, 300-326
$\frac{3}{4} = \{\frac{1}{2}|1\}$, three-quarters, 17
$+_{\mathbf{on}} = \{0|\mathbf{oof}\} = \{0||0|\mathbf{off}\} = \mathbf{tiny}$, 353, 357
$+_{\frac{1}{4}} = \{0 \,||\, 0|-\frac{1}{4}\}$, tiny-a-quarter, 126
$+_2 = \{0 \,||\, 0|-2\}$, tiny-two, 126
$+_x = \{0 \,||\, 0|-x\}$, tiny-x, 126
$\mathbf{tis} = \{\mathbf{tisn}|\ \ \} = 1\&0$, 342, 374
$\mathbf{tisn} = \{\ \ |\mathbf{tis}\} = 0\&1$, 342, 374
$(l, r)(l, r)_c$, Toads-and-Frogs positions, 127, 136, 368, 375, 376
$= \{\Downarrow *|0\} = 3.\downarrow$, treble-down, 71, 347
$= \{0|\Uparrow *\} = 3.\uparrow$, treble-up, 71, 347
$3♠ = \{0|\mathbf{deuce}\} = \mathbf{trey} = \mathbf{ace} + \mathbf{deuce}$, 357
$\triangle$, triangular number, 253
$2 = \{1|\ \ \}$, two, 7, 19
$\mathbf{under} = \{\mathbf{under}|0\} = -\mathbf{over}$, 341
$\vee$**.union**, or, 300
$\uparrow = \{0|*\}$, up, 64, 71, 261
$\uparrow^\alpha$, up-alpha, 341
$\uparrow^2 = \{0|\downarrow *\}$, up-second, 235, 236
$\uparrow * = \{0, *|0\}$, up-star, 65, 229
$\uparrow^3 = \{0|\downarrow + \downarrow_2 + *\}$, up-third, 235, 236
$\uparrow_{abc\ldots}$, 261
$\mathbf{upon} = \{\mathbf{upon}|*\}$, 341, 375
$\mathbf{upon}* = \{0, \mathbf{upon} * |0\}$, 341
$\dotplus$, upsum, 336, 357
$\bigtriangledown$, ur, urgent number, 312
$\mathbb{Z} = \{\ldots, -2, -1, 0, 1, 2, \ldots\}$, the set of integers, 334
$0 = \{\ \ |\ \ \} = 0* = 0.\uparrow$, zero, 7, 41

Index

cooling, 151, 152, 154, 167, 179, 231
  formula, 151